CAMBRIDGE TRACTS IN MATHEMATICS

214 The Mathieu Groups

CAMBRIDGE TRACTS IN MATHEMATICS

A complete list of books in the series can be found at www.cambridge.org/mathematics. Recent titles include the following:

181. Totally Positive Matrices. By A. PINKUS
182. Nonlinear Markov Processes and Kinetic Equations. By V. N. KOLOKOLTSOV
183. Period Domains over Finite and p-adic Fields. By J.-F. DAT, S. ORLIK, and M. RAPOPORT
184. Algebraic Theories. By J. ADÁMEK, J. ROSICKÝ, and E. M. VITALE
185. Rigidity in Higher Rank Abelian Group Actions I: Introduction and Cocycle Problem. By A. KATOK and V. NIȚICĂ
186. Dimensions, Embeddings, and Attractors. By J. C. ROBINSON
187. Convexity: An Analytic Viewpoint. By B. SIMON
188. Modern Approaches to the Invariant Subspace Problem. By I. CHALENDAR and J. R. PARTINGTON
189. Nonlinear Perron–Frobenius Theory. By B. LEMMENS and R. NUSSBAUM
190. Jordan Structures in Geometry and Analysis. By C.-H. CHU
191. Malliavin Calculus for Lévy Processes and Infinite-Dimensional Brownian Motion. By H. OSSWALD
192. Normal Approximations with Malliavin Calculus. By I. NOURDIN and G. PECCATI
193. Distribution Modulo One and Diophantine Approximation. By Y. BUGEAUD
194. Mathematics of Two-Dimensional Turbulence. By S. KUKSIN and A. SHIRIKYAN
195. A Universal Construction for Groups Acting Freely on Real Trees. By I. CHISWELL and T. MÜLLER
196. The Theory of Hardy's Z-Function. By A. IVIĆ
197. Induced Representations of Locally Compact Groups. By E. KANIUTH and K. F. TAYLOR
198. Topics in Critical Point Theory. By K. PERERA and M. SCHECHTER
199. Combinatorics of Minuscule Representations. By R. M. GREEN
200. Singularities of the Minimal Model Program. By J. KOLLÁR
201. Coherence in Three-Dimensional Category Theory. By N. GURSKI
202. Canonical Ramsey Theory on Polish Spaces. By V. KANOVEI, M. SABOK, and J. ZAPLETAL
203. A Primer on the Dirichlet Space. By O. EL-FALLAH, K. KELLAY, J. MASHREGHI, and T. RANSFORD
204. Group Cohomology and Algebraic Cycles. By B. TOTARO
205. Ridge Functions. By A. PINKUS
206. Probability on Real Lie Algebras. By U. FRANZ and N. PRIVAULT
207. Auxiliary Polynomials in Number Theory. By D. MASSER
208. Representations of Elementary Abelian p-Groups and Vector Bundles. By D. J. BENSON
209. Non-homogeneous Random Walks. By M. MENSHIKOV, S. POPOV and A. WADE
210. Fourier Integrals in Classical Analysis (Second Edition). By C. D. SOGGE
211. Eigenvalues, Multiplicities and Graphs. By C. R. JOHNSON and C. M. SAIAGO
212. Applications of Diophantine Approximation to Integral Points and Transcendence. By P. CORVAJA and U. ZANNIER
213. Variations on a Theme of Borel. By S. WEINBERGER
214. The Mathieu Groups. By A. A. IVANOV
215. Slenderness I: Abelian Categories. By RADOSLAV DIMITRIC

The Mathieu Groups

A. A. IVANOV
Imperial College London
and
Institute for System Analysis FRC, CSC, RAS

CAMBRIDGE
UNIVERSITY PRESS

CAMBRIDGE
UNIVERSITY PRESS

University Printing House, Cambridge CB2 8BS, United Kingdom

One Liberty Plaza, 20th Floor, New York, NY 10006, USA

477 Williamstown Road, Port Melbourne, VIC 3207, Australia

314-321, 3rd Floor, Plot 3, Splendor Forum, Jasola District Centre, New Delhi - 110025, India

79 Anson Road, #06-04/06, Singapore 079906

Cambridge University Press is part of the University of Cambridge.

It furthers the University's mission by disseminating knowledge in the pursuit of education, learning and research at the highest international levels of excellence.

www.cambridge.org
Information on this title: www.cambridge.org/9781108429788
DOI: 10.1017/9781108555289

First published 2018

A catalogue record for this publication is available from the British Library

Library of Congress Cataloging in Publication data
Names: Ivanov, A. A. (Aleksandr Anatolievich), 1958– author.
Title: The Mathieu groups / A.A. Ivanov (Imperial College London).
Description: Cambridge ; New York, NY : Cambridge University Press, [2018] |
Series: Cambridge tracts in mathematics
Identifiers: LCCN 2018011064 | ISBN 9781108429788
Subjects: LCSH: Mathieu groups. | Permutation groups. | Finite groups.
Classification: LCC QA171 .I93 2018 | DDC 512/.23–dc23
LC record available at https://lccn.loc.gov/2018011064

ISBN 978-1-108-42978-8 Hardback

In gratitude to Ernie Shult

Contents

Preface

"There are almost as many different constructions of M_{24} as there have been mathematicians interested in that most remarkable of all finite groups".[1]

In this book the study of the Mathieu group M_{24} (and other Mathieu groups it contains) falls within the scope of what E. E. Shult[2] called the Ivanov–Shpectorov theory of geometries. This theory has been developed to construct and identify large sporadic simple groups including the Baby Monster,[3,4] the Fourth Janko Group J_4[5] and the Monster.[6] The most dramatic outcome of the theory was the proof of the famous Y-presentation conjecture for the Monster, which for a long time remained unobtainable by use of the other techniques.[7] In the case of M_{24} the way in which the theory develops can be projected onto the familiar structures of the Steiner system on 24 points and the Golay code, thus presenting a bold illustration of the theory as well as providing a fresh look at familiar, nearly classical structures. I am extremely grateful to Madeleine Whybrow, William Giuliano and the anonymous referees for suggesting thoughtful corrections, clarifications and modifications after reading earlier versions of the book.

[1] J. H. Conway, The Golay Codes and the Mathieu Groups, in *Sphere Packings, Lattices and Groups*, ed. J. Conway and N.J.A Sloane, Springer, New York, 1988, pp. 299–330.

[2] E. E. Shult, lecture notes on coverings of graphs, Kansas State University, 1997.

[3] A. A. Ivanov, *Geometry of Sporadic Groups* I, Cambridge University Press, Cambridge, 1999.

[4] A. A. Ivanov and S. V. Shpectorov, *Geometry of Sporadic Groups* II, Cambridge University Press, Cambridge, 2002

[5] A. A. Ivanov, J_4, Oxford University Press, Oxford, 2004.

[6] A. A. Ivanov, *The Monster Group and Majorana Involutions*, Cambridge University Press, Cambridge, 2009.

[7] A. A. Ivanov, Y-groups via transitive extensions, *J. Algebra* **218** (1999), 412–435.

1

The Mathieu Group M_{24} As We Knew It

In this chapter we start by reviewing the common way to describe the Mathieu group $G = M_{24}$ as the automorphism group of the binary Golay code and of the Steiner system formed by the minimal codewords in the Golay code. This discussion will lead us to the structure of the octad–trio–sextet stabilizers $\{G_1, G_2, G_3\}$. The starting point is the observation that, when forming a similar triple $\{H_1, H_2, H_3\}$ comprised of the stabilizers of one-, two- and three-dimensional subspaces in $H = L_5(2)$, we have

$$G_1 \cong H_1 \cong 2^4 : L_4(2), \quad G_2 \cong H_2 \cong 2^6 : (L_3(2) \times S_3),$$
$$[G_1 : G_1 \cap G_2] = [H_1 : H_1 \cap H_2] = 15,$$
$$[G_2 : G_1 \cap G_2] = [H_2 : H_1 \cap H_1] = 3,$$

so that the *amalgams* $\{G_1, G_2\}$ and $\{H_1, H_2\}$ have the *same type* according to Goldschmidt's terminology, although they are not isomorphic and differ by a *twist* performed by an outer automorphism of

$$H_1 \cap H_2 \cong G_1 \cap G_2 \cong (2^3 \times 2^3) : (L_3(2) \times 2),$$

which permutes conjugacy classes of $L_3(2)$-subgroups. This observation led us to the construction of M_{24} as the universal completion of a twisted $L_5(2)$-amalgam.

1.1 The Golay Code

Commonly the Mathieu group M_{24} is defined as the automorphism group of the Golay code, which by definition is a (a) binary (b) linear code (c) of length 24, which is (d) even, (e) self-dual and (f) has no codewords of weight 4.

By (a) we can identify a codeword with its support in the standard basis and view the Golay code as a pair $(\mathcal{P}, \mathcal{C})$, where $\mathcal{P}$ is a set and $\mathcal{C}$ is a collection of subsets of $\mathcal{P}$, so that $\mathcal{P}$ together with an ordering can be identified with the standard basis of the vector space $2^{\mathcal{P}} := \{A \mid A \subseteq \mathcal{P}\}$ in which addition is performed by the symmetric difference operator:

$$A + B := (A \cup B) \setminus (A \cap B).$$

The *weight* of $A \subseteq \mathcal{P}$ is its cardinality $|A|$ and the remaining defining properties of the Golay code can be restated as follows:

(b) $\mathcal{C}$ is closed under addition;
(c) $|\mathcal{P}| = 24$;
(d) every subset in $\mathcal{C}$ has an even number of elements;
(e) a subset B of $\mathcal{P}$ intersects evenly every $A \in \mathcal{C}$ exactly when B is taken from $\mathcal{C}$;
(f) the minimal weight of $\mathcal{C}$ is 8.

It is convenient to deduce the numerology of the Golay code starting with consideration of its co-code $\mathcal{C}^* := 2^{\mathcal{P}}/\mathcal{C}$. For $A \subseteq \mathcal{P}$ let A^* denote the image of A in the co-code: $A^* = A + \mathcal{C}$. The following *cardinality* attribute is immediately evident from the minimal weight of $\mathcal{C}$ being 8.

Lemma 1.1 *Whenever two distinct subsets A and B in $\mathcal{P}$, each of cardinality at most 4, have the same image in $\mathcal{C}^*$, the equality $|A| = |B| = 4$ holds, and A is disjoint from B.* □

Let $\mathcal{P}_{(4)}$ be the set of all subsets of cardinality at most 4 in $\mathcal{P}$. Since it is not possible to find more than six pairwise disjoint 4-subsets in a 24-set, the *cardinality* attribute Lemma 1.1 gives the following lower bound on the size of the image $\mathcal{P}^*_{(4)}$ of $\mathcal{P}_{(4)}$ in $\mathcal{C}^*$:

$$|\mathcal{P}^*_{(4)}| \geq 1 + 24 + \binom{24}{2} + \binom{24}{3} + \frac{1}{6}\binom{24}{4} = 2^{12}.$$

On the other hand, the self-duality of $\mathcal{C}$ means that it is a maximal isotropic subspace in $2^{\mathcal{P}}$ with respect to the non-degenerate symplectic form

$$f : (A, B) \mapsto |A \cap B| \mod 2.$$

Thus the dimension of $\mathcal{C}$ is exactly *half* the dimension of $2^{\mathcal{P}}$, so that $|\mathcal{C}| = |\mathcal{C}^*| = 2^{12}$, and the above lower bound is attained. Therefore $\mathcal{C}^* = \mathcal{P}^*_{(4)}$, which brings about the following *representative* principle.

Lemma 1.2 *For every $X \subseteq \mathcal{P}$ the coset X^* contains a subset A of cardinality at most 4. The cardinality of such a subset A is uniquely determined by X. If the cardinality of A is strictly less than 4 then A itself is uniquely determined by*

X, but if the cardinality is 4 then there are precisely six choices for A forming a partition of $\mathcal{P}$ into six disjoint 4-subsets. □

Let us introduce some further terminology by calling the minimal non-empty subsets of $\mathcal{C}$ (having size eight) *octads*, and denoting by $\mathcal{B}$ the set of octads. A partition of $\mathcal{P}$ into six 4-subsets such that the union of any two of them is an octad is known as a *sextet*. Notice that the partition which appeared in the last sentence of the *representative* principle Lemma 1.2 is a sextet. A partition of $\mathcal{P}$ into three disjoint octads is called a *trio*. Let $\mathcal{S}$ and $\mathcal{T}$ denote the sets of sextets and trios, respectively. Given a sextet, one can easily construct a trio by taking a sub-partition of the sextet, and then $\mathcal{P}$ is an element of $\mathcal{C}$, since it is the sum of the octads in a trio. Also two distinct octads cannot share a 5-set, because of the minimal weight property. These observations, together with the *representative* principle, give the following *Steiner* attribute.

Lemma 1.3 *Every 4-subset of $\mathcal{P}$ is a member of a unique sextet, and every 5-subset is contained in a unique octad. Furthermore, trios exist and $\mathcal{P}$, viewed as an element from $2^{\mathcal{P}}$, is contained in $\mathcal{C}$.* □

Now easy combinatorial counting demonstrates the equalities

$$|\mathcal{S}| = \frac{1}{6}\binom{24}{4} = 1771 \text{ and } |\mathcal{B}| = \frac{\binom{24}{5}}{\binom{8}{5}} = 759.$$

The pair $(\mathcal{P}, \mathcal{B})$ is a Steiner system of type $S(24, 8, 5)$, which by definition is a collection of 8-subsets in a 24-set such that every 5-subset is covered by exactly one 8-subset from the collection.

In order to calculate the number of trios we first analyze how two octads can intersect. Let n_i denote the number of octads intersecting a given octad B in exactly i elements. Clearly $n_8 = 1$, while $n_i = 0$ whenever i is odd by the self-duality condition, and $n_6 = 0$ by the minimal weight condition. A 4-subset X is contained in five octads which are unions of the pairs of 4-subsets in the sextet determined by X and B is one of them. This readily gives the equality

$$n_4 = \binom{8}{4}(5 - 1) = 280.$$

In order to calculate n_2 we consider a 5-subset Q which intersects B in exactly two elements. The unique octad $B(Q)$ containing Q might intersect B in four points, but all such subsets Q can be counted, since n_4 is known. For the remaining Qs we have $|B \cap B(Q)| = 2$, which gives

$$n_2 = \left[\binom{8}{2}\binom{16}{3} - n_4\binom{4}{2}\binom{4}{3}\right] \Big/ \binom{6}{3} = 448.$$

Finally,

$$n_0 = |\mathcal{B}| - n_8 - n_4 - n_2 = 30.$$

Since $\mathcal{C}$ is linear and contains $\mathcal{P}$, the complement of the union of two disjoint octads is again an octad, so that an octad is contained in $15 = n_0/2$ trios and

$$|\mathcal{T}| = |\mathcal{B}| \cdot n_0/(2 \cdot 3) = 3795.$$

Any two disjoint octads determine a trio, which implies that any two octads that are disjoint from a given octad B are either disjoint forming a trio with B, or intersect each other in a 4-subset. Since an octad is contained in more than one trio, this demonstrates that every trio is sub-partitioned by a sextet. There are 15 partitions of a six-element set into three disjoint pairs, therefore every sextet sub-partitions 15 trios. In view of the equality

$$15 \cdot |\mathcal{S}| = 7 \cdot |\mathcal{T}|,$$

this implies that every trio is sub-partitioned by exactly seven sextets.

1.2 The Octad Graph

Define the *octad graph* Γ to be the graph having the set $\mathcal{B}$ of octads as the vertex set, in which two vertices are adjacent whenever they are disjoint as octads. From the discussion in the previous section, we have that Γ is regular of valency $n_0 = 30$ and every edge is contained in a unique triangle corresponding to a trio, and hence there are 15 triangles through a given vertex.

Theorem 1.4 *The octad graph Γ is distance-regular with the following intersection diagram:*

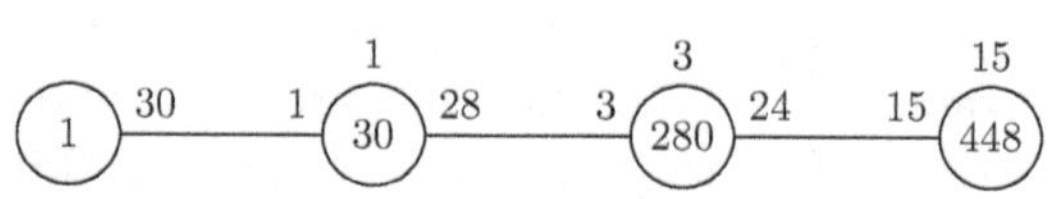

The intersection diagram indicates that the diameter of Γ is 3, while the number of vertices at distance i from any given vertex-octad B is 1, 30, 280 and 448 for $i = 0, 1, 2$ and 3, respectively. These numbers appear inside the circles read from left to right. If B' is a vertex at distance i from B then, among the 30 vertices adjacent to B', exactly c_i vertices are at distance $i - 1$ from B, a_i at distance i and b_i at distance $i + 1$. The parameters c_i, a_i and b_i depend not on the individual choices of B and B', but only on the distance between them, and on the diagram they are drawn around the ith circle at 9, 12 and 3 o'clock, respectively.

Proof of Theorem 1.4 The values $b_0 = n_0 = 30$, $c_1 = 1$, $a_1 = 1$ and $b_1 = n_0 - c_1 - a_1 = 24$ have already been justified. For a sextet S the subgraph induced by the vertices or octads which are unions of pairs of 4-subsets in S is a 15-vertex subgraph of valency 6. This subgraph is the graph on 2-subsets of a six-element set in which two subsets are adjacent whenever they are disjoint. These subgraphs will be called *quads*, with every quad being distance-regular as described by the following diagram:

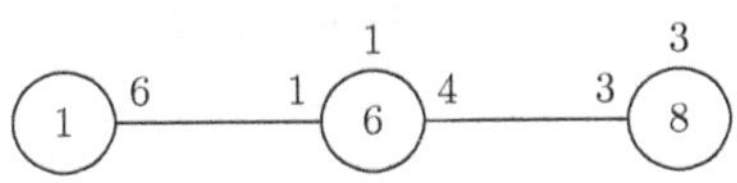

Since any two octads intersecting in four elements are unions of pairs of 4-subsets in the sextet determined by the intersection, we conclude that such octads are at distance 2 in Γ and are contained in a common quad. The diagram of the quad shows that $c_2 \geq 3$ and $a_2 \geq 3$, and hence $b_2 \leq 24$. Notice that at this stage the parameters c_2, a_2 and b_2 might still depend on the particular choice of the pair of vertices at distance 2 in Γ, but for any vertex of Γ the number $N_{2,3}$ of edges joining vertices at distance 2 with vertices at distance 3 from that vertex satisfies the inequality

$$N_{2,3} \leq 280 \cdot 24.$$

Suppose that B and B' are octads intersecting in a 2-subset, and let $T = \{B, B_1, B_2\}$ be a trio containing B. Then the 6-set $B' \setminus B$ splits between B_2 and B_3. Since any two octads intersect evenly and never share a 6-set, up to reordering the splitting is $6 = 4+2$. Therefore, in every triangle-trio containing B there is exactly one vertex at distance 2 from B' and the other two vertices, including B, are at distance 3, so that $c_3 = a_3 = 15$ and

$$N_{2,3} = 448 \cdot 15.$$

Since the value of $N_{2,3}$ attains the above upper bound, the equality $c_2 = a_2 = 3$ holds and the proof of distance-regularity is complete. □

1.3 A Review

In this section we reveal the existence and uniqueness features of the Golay code $\mathcal{C}$ and discuss the automorphism group of $\mathcal{C}$ along with some of its important subgroups. In all the uniqueness statements the caveat *up to isomorphism* is implicit. The following statement of *existence-uniqueness* was first established by E. Witt in 1938 for the Steiner system.[1] The construction of the Golay code in 1949[2] extends it to the code, and for the octad graph the uniqueness was proved by A. E. Brouwer.[3]

Theorem 1.5 *There exists exactly one Steiner system* $(\mathcal{P}, \mathcal{B})$ *of type* $S(24, 8, 5)$. *The span of* $\mathcal{B}$ *in the power space* $2^{\mathcal{P}}$ *is the unique Golay code, and the octad graph* Γ *is the unique distance-regular graph with its intersection diagram subject to the existence of the quads.* □

The Steiner system $(\mathcal{P}, \mathcal{B})$ is uniquely reconstructible from the Golay code $\mathcal{C}$. In fact $\mathcal{C}$ can be recovered from the octad graph Γ, although we postpone the explanation of this procedure. In any event, all of the three objects $(\mathcal{P}, \mathcal{B})$, $\mathcal{C}$ and Γ have the same automorphism group G. The following 5-*transitivity* attribute is the central point of the review. This and further theorem attributes in this section will be proved later in the book within our construction of the Mathieu group M_{24} by group amalgams (see Theorem 2.2 and sections after that theorem). Other proofs can be found in M. Aschbacher's book,[4] in Chapter 6 of the book by J. D. Dixon and B. Mortimer,[5] in my book[6] and in many other books and journal articles.

Theorem 1.6 *The automorphism group* G *of* $(\mathcal{P}, \mathcal{B})$, $\mathcal{C}$ *and* Γ *is the Mathieu group* M_{24} *discovered by É. Mathieu in 1873.*[7] *It is simple of order*

$$|M_{24}| = 2^{10} \cdot 3^3 \cdot 5 \cdot 7 \cdot 11 \cdot 23$$

and acts 5-fold transitively on the 24-set $\mathcal{P}$. □

The 5-fold transitivity is by definition the transitivity of G on the set of ordered 5-subsets of $\mathcal{P}$. The following *flag-transitivity* attribute is the foundation of our treatment of G. Here a *flag* $\Phi = \{B, T, S\}$ is an octad–trio–sextet triple such that S sub-partitions T, and B is one of the three octads in T.

[1] E. Witt, Über Steinersche Systeme, *Abh. Math. Seminar Hamburg* **12** (1938), 265–275.

[2] M. J. E. Golay, Notes on digital coding, *Proc. IRE* **37** (1949), 657.

[3] A. E. Brouwer, The uniqueness of the near hexagon on 759 points, in *Finite Geometries*, ed. N. L. Johnson, M. J. Kallaher and C. T. Long, Marcel Dekker, New York, 1982, pp. 47–60.

[4] M. Aschbacher, *Sporadic Groups*, Cambridge University Press, Cambridge, 1994.

[5] J. D. Dixon and B. Mortimer, *Permutation Groups*, Springer, Berlin, 1996.

[6] A. A. Ivanov, *Geometry of Sporadic Groups* I, Cambridge University Press, Cambridge, 1999.

[7] É. Mathieu, Sur la fonction cinq fois transitive de 24 quantités, *J. Math. Pures Appl.* **18** (1873), 25–46.

Theorem 1.7 *The group G acts transitively on the set of flags Φ, the stabilizer of Φ has order $2^{10} \cdot 3$, and thus contains a Sylow 2-subgroup of G as a subgroup of index 3.* □

With the flag $\Phi = \{B, T, S\}$ as above, let G_1, G_2 and G_3 be the stabilizers in G of B, T and S, respectively. Then we have the following *parabolic structure* attribute formulated in the standard group-theoretical terms.

Theorem 1.8 *The following isomorphisms hold:*

$$G_1 \cong 2^4 : L_4(2), \quad G_2 \cong 2^6 : (L_3(2) \times S_3), \quad G_3 \cong 2^6 : 3 \cdot S_6. \qquad \square$$

By the *flag-transitivity* attribute, G acts transitively on each of the sets $\mathcal{B}$, $\mathcal{T}$ and $\mathcal{S}$ as on the cosets of G_1, G_2 and G_3, respectively. In particular, the action of G on Γ is vertex-transitive. Furthermore, the following *distance-transitivity* property holds.

Theorem 1.9 *The action of M_{24} on the octad graph Γ is distance-transitive, and so it is vertex-transitive, and G_1 permutes transitively the vertices at distance i from B for every $i = 0, 1, 2$ and 3.* □

The following *simple connectedness* attribute proved by M. Ronan[8] assures the success of our construction of the Mathieu group M_{24} as the universal completion of the Mathieu amalgam $\{G_1, G_2, G_3\}$.

Theorem 1.10 *The octad–trio–sextet geometry of the Mathieu group M_{24} is simply connected.* □

[8] M. A. Ronan, Locally truncated buildings and M_{24}, *Math. Z.* **180** (1982), 469–501.

2

The Amalgam Method

In this chapter we introduce the notion of an amalgam, discuss its role in modern algebraic combinatorics and group theory and define the Mathieu amalgam, which will be the principal object in our construction of the Mathieu group M_{24} and investigation of its structure, subgroups and representations.

2.1 Amalgams

In this section, for a group A, it is convenient to use the full name $(A, *)$, where A is the element set and $*$ is the group product on A. An (abstract) *amalgam* of rank n is a collection

$$\mathcal{A} = \{(A_i, *_i) \mid 1 \leq i \leq n\}$$

of n groups (called *members* of the amalgam) whose products coincide on the intersections of the element sets: whenever $a, b \in A_i \cap A_j$ the equality $a *_i b = a *_j b$ holds. A *concrete* amalgam is a collection of subgroups in a group where the group products in members are the restrictions of the product in the whole group. The *core* of an amalgam is the largest subgroup contained in each of its members as a normal subgroup.

The notions of homomorphism, isomorphism and automorphism in the class of group amalgams have the obvious meanings. Groups can be viewed as rank-1 amalgams and an amalgam B is a subamalgam in an amalgam A whenever every member of B is a subgroup in a member of A. A *completion* of an amalgam is a realization of (a homomorphic image of) the amalgam as a concrete amalgam. Thus a completion of the amalgam $\mathcal{A}$ as above is a group $(A, \ \cdot\ )$ together with a mapping ψ of the element set of $\mathcal{A}$ into A, such that the restriction of ψ to every member of $\mathcal{A}$ is a group homomorphism:

$$\psi(a *_i b) = \psi(a) \cdot \psi(b),$$

for all $a, b \in A_i$ and $1 \leq i \leq n$. Such a completion is said to be *faithful* if ψ is injective, and *generating* if A is generated by the image of ψ. Every amalgam possesses a unique *universal completion group* which has the element set of the amalgam as the set of generators and where the relations are all the equalities which hold in the members of the amalgam. The universality property means that every completion is the universal one followed by a group homomorphism.

If $\psi : \mathcal{A} \to A$ is a generating completion then the associated *coset graph* Δ is an n-partite graph, whose ith part is comprised by the right cosets of $\psi(A_i)$ in A:

$$\Delta = \Delta^{(1)} \cup \Delta^{(2)} \cup \cdots \cup \Delta^{(n)}, \quad \Delta^{(i)} = \{\psi(A_i)a \mid a \in A\}.$$

Two vertex-cosets are adjacent in Δ whenever they are from different parts and have a non-empty intersection. Notice that, whenever $\psi(A_i)a$ properly intersects $\psi(A_j)b$, the intersection is a coset of $\psi(A_i \cap A_j)$. The group A acts naturally on Δ, preserving the parts together with the adjacency relation. The core of the amalgam is always in the kernel of the action. A homomorphism of two faithful generating completions induces a *local* morphism of the corresponding coset graphs, where local means that the structure of the subgraphs induced by the neighbours of every vertex stays unchanged. The following fundamental result was proved independently by A. Pasini,[1] S. V. Shpectorov[2] and J. Tits.[3]

Theorem 2.1 *The coset graph of a generating completion is simply connected with respect to the local morphisms if and only if the completion is universal.* □

If a cycle in the coset graph is called *local* whenever it is contained in the neighbourhood of a vertex, then the above assertion means that the fundamendal group of the coset graph of the universal completion of an amalgam is generated by the homotopy classes of the local cycles.

A *presentation* of an amalgam $\mathcal{A} = \{(A_i, *_i) \mid 1 \leq i \leq n\}$ is a pair (X, R) consisting of a subset

$$X \subseteq \bigcup_{i=1}^{n} A_i$$

[1] A. Pasini, Some remarks on covers and apartments, in *Finite Geometries*, ed. C. A. Baker and L. M. Batten, Marcel Dekker, New York, 1985, 233–250.

[2] S. V. Shpectorov, unpublished manuscript dating from c. 1985.

[3] J. Tits, Ensembles ordonnés, immeubles et sommes amalgamées, *Bull. Soc. Math. Belg.* **A38** (1986), 367–387.

of the element set of the amalgam together with a set of relations R which involve elements from X, and which satisfy the following conditions:

(i) for every $1 \leq i \leq n$ the set $X \cap A_i$ generates A_i;
(ii) for all $1 \leq i < j \leq n$ the set $X \cap A_i \cap A_j$ generates $A_i \cap A_j$;
(iii) if R_i is the set of relations in R which involve only elements from A_i, then $R = \cup_{i=1}^{n} R_i$ and $(X \cap A_i, R_i)$ is a presentation for A_i for every $1 \leq i \leq n$.

For an amalgam $\mathcal{A}$ whose members are finite groups, the *universal* presentation is the pair $(\cup_{1=1}^{n} A_i, R^{(u)})$, where $R^{(\mathrm{u})}$ is the set of all equalities $abc^{-1} = 1$, such that $a, b \in A_i$ for some $1 \leq i \leq n$ and $c = a *_i b$. It is clear that a presentation of an amalgam $\mathcal{A}$ (and particularly its universal presentation) is a presentation for the universal completion group of $\mathcal{A}$ in the usual sense.

2.2 Goldschmidt's Amalgams

Rank-2 amalgams are usually treated in a slightly more refined, although equivalent, setting. Such an amalgam $\mathcal{A} = \{A_1, A_2\}$ is defined as a triple of groups A_1, A_2 and A_{12} together with a pair of injective homomorphisms

$$\varphi_1 \colon A_{12} \to A_1, \quad \varphi_2 \colon A_{12} \to A_2.$$

Then the *type* of $\mathcal{A}$ is a specification of the isomorphism types of A_1, A_2, A_{12} and of the subgroups $\varphi_1(A_{12})$ and $\varphi_2(A_{12})$ in A_1 and A_2, respectively, up to conjugation in the automorphism groups of A_1 and A_2.

Goldschmidt's lemma *Let $\mathcal{A} = (A_1, A_2, A_{12}, \varphi_1, \varphi_2)$ be a rank-2 amalgam. Let $B = \mathrm{Aut}\,(A_{12})$, let B_1 be the natural image in B of the normalizer in* $\mathrm{Aut}\,(A_1)$ *of $\varphi(A_{12})$, and let B_2 be defined similarly with respect to A_2. Then every amalgam having the same type as $\mathcal{A}$ is isomorphic to*

$$\mathcal{A}_\alpha = (A_1, A_2, A_{12}, \varphi_1, \varphi_2\alpha)$$

for some $\alpha \in B$. Furthermore, $\mathcal{A}_\alpha$ and $\mathcal{A}_\beta$ are isomorphic if and only if

$$B_1\alpha B_2 = B_1\beta B_2.$$

The main result of Goldschmidt's groundbreaking paper[4] was the list of all core-free rank-2 amalgams (P_1, P_2, P_{12}) with $[P_1 : P_{12}] = [P_2 : P_{12}] = 3$, which is reproduced in Table 2.1 (with some notations adjusted to the current exposition).

[4] D. M. Goldschmidt, Automorphisms of trivalent graphs, *Ann. Math.* **111** (1980), 377–406.

Table 2.1

Amalgams	(P_1, P_2)	Faithful finite completions
G_1	(Z_3, Z_3)	$Z_3 \times Z_3$
G_1^1	(S_3, S_3)	$(Z_3 \times Z_3).Z_2$
G_1^2	(S_3, Z_6)	$(Z_3 \times Z_3).Z_2$
G_1^3	(D_{12}, D_{12})	$L_2(11)$
G_2	(D_{12}, A_4)	$L_2(11)$
G_2^1	(D_{24}, S_4)	$L_2(23)$
G_2^2	$(D_8 : S_3, S_4)$	A_7
G_2^3	$(D_{12} \times 2, A_4 \times 2)$	$S_3 \wr Z_3$
G_2^4	$(D_8 \times S_3, S_4 \times 2)$	S_7
G_3	(S_4, S_4)	$L_3(2)$
G_3^1	$(S_4 \times 2, S_4 \times 2)$	S_6
G_4	$((Q_8 * Z_4).S_3, (Z_4 \times Z_4).S_3)$	$U_3(3)$
G_4^1	$((Q_8 * Q_8)^1\, S_3, (Z_4 \times Z_4).D_{12})$	$G_2(2)$
G_5	$((Q_8 * Q_8)^2.S_3, (Z_4 \times Z_4).D_{12})$	M_{12}
G_5^1	$((Q_8 * Q_8)^2.D_{12}, (Z_4 \times Z_4).(S_3 : D_8))$	$\mathrm{Aut}(M_{12})$

The third column in Table 2.1 gives sample completions (there is, of course, nothing unique about these completions). For the last three amalgams, $(Q_8 * Q_8)^n$ is a semidirect product with n non-central 2-chief factors ($n = 1, 2$) and $(Q_8 * Q_8)^2.D_{12}$ is a non-split extension. Thus in P_1 of G_4^1 but not of G_5 one Q_8-central factor is centralized by an element of order 3 from S_3.

Although the method of group amalgams is probably the most powerful tool in the modern study of groups, the theory of amalgams is pretty much in the making and the present book contributes to building up the theory. The rank-2 amalgams are better understood:[5,6] the universal completion of such an amalgam is the free product of its members amalgamated over the intersection and the coset graph is a tree on which the universal completion acts as a locally finite edge-transitive automorphism group (we exclude the trivial case in which one member is a subgroup of the other one). Thus one needs the rank to be at least 3 in order to obtain a finite group as a universal completion.[7,8] There is

[5] J. P. Serre, *Trees*, Springer, New York, 1980.

[6] A. Delgado, D. Goldschmidt and B. Stellmacher, *Groups and Graphs: New Results and Methods*, Birkhäuser, Basel, 1985.

[7] M. Aschbacher and Y. Segev, Extending morphisms of groups and graphs, *Ann. Math.* **135** (1992), 297–323.

[8] A. A. Ivanov and S. V. Shpectorov, Applications of group amalgams to algebraic graph theory, in *Investigation in Algebraic Theory of Combinatorial Objects*, ed. I. A.. Faradzev, A. A. Ivanov, M. H. Klin and A. J. Woldar, Kluwer, Dordrecht, 1994, pp. 417–442.

no direct analogy of Goldschmidt's lemma for higher ranks and the following section makes an attempt at generalization in the special case of so-called *constrained amalgams*, which are the most important type in applications.

2.3 Constrained Amalgams

Goldschmidt's lemma provides a tool for classifying rank-2 amalgams. Since the universal completion of such an amalgam is a free amalgamated product, which is infinite (unless the whole amalgam sits in one of its members), we need the rank to be at least 3 in order to find some finite groups as universal completions. The general question about isomorphism types of rank-3 amalgams is more subtle in general, so we restrict ourselves to a special class.

A rank-3 amalgam $\mathcal{A} = \{A_1, A_2, A_3\}$ is said to be *constrained* if there is a subgroup X_3 in $A_1 \cap A_2$ such that the following conditions hold, where $A_{ij} = A_i \cap A_j$ for $1 \leq i < j \leq 3$:

(C1) A_3 contains X_3 as a normal subgroup;
(C2) $A_{i3} = N_{A_i}(X_3)$ for $i = 1$ and 2;
(C3) A_3 is generated by A_{13} and A_{23};
(C4) $C_{A_3}(X_3) = Z(X_3)$.

Theorem 2.2 *The isomorphism type of an amalgam $\mathcal{A} = \{A_1, A_2, A_3\}$ satisfying (C1) to (C4) is uniquely determined by the isomorphism type of the subamalgam $\{A_1, A_2\}$ and the choice of a U_3-invariant complement K_3 to $Z(X_3)$ in the centralizer of X_3 in the universal completion U_3 of the amalgam*

$$\mathcal{A}_3 = \{A_{13} = N_{A_1}(X_3), A_{23} = N_{A_2}(X_3)\}.$$

In particular, if $Z(X_3) = 1$ then the isomorphism type of $\mathcal{A}$ is uniquely determined by that of $\{A_1, A_2\}$.

Proof Suppose that the isomorphism type of the amalgam $\{A_1, A_2\}$ is fixed along with the subgroup X_3 and we are faced with the problem of describing up to isomorphism the amalgam $\mathcal{A} = \{A_1, A_2, A_3\}$ satisfying (C1) to (C4). By (C1) to (C3), the subamalgam $\mathcal{A}_3 = \{A_{13}, A_{23}\}$ is uniquely determined in the subamalgam $\{A_1, A_2\}$ and $\mathcal{A}_3$ generates A_3. Therefore A_3 is a quotient of the universal completion U_3 of $\mathcal{A}_3$ and the isomorphism type of $\mathcal{A}$ is uniquely determined by the kernel K_3 of the corresponding homomorphism

$$\psi : U_3 \to A_3.$$

Both K_3 and X_3 are normal subgroups in U_3, and, since X_3 maps isomorphically into A_3, they have a trivial intersection and hence must commute, so that $K_3 \leq C_{U_3}(X_3)$. Clearly $C_{X_3}(X_3) = Z(X_3)$. By (C4), this means that K_3 is a U_3-invariant complement to $Z(X_3)$ in $C_{U_3}(X_3)$. It is clear that every choice of such a complement leads to an amalgam $\mathcal{A}$ with the required properties. □

2.4 Understanding Completions

Now suppose that an amalgam

$$\mathcal{A} = \{A_1, A_2, A_3\}$$

(which might or might not be constrained) is constructed or just given and we are faced with the question of whether it possesses a faithful completion and, if so, what the universal completion would be. This question is potentially very hard due to the non-solvability of the word problem in a group, but we hope that the group we are after is somewhat special. The methods developed so far can be specified as follows.

Coset enumeration. If the target group is not too large one might write down a presentation of the amalgam and run a coset enumeration procedure on a computer to determine the order of the universal completion. In this book we apply this procedure in Chapter 7 to identify M_{24} and He as the universal completion of the relative amalgams. Alternatively one could try to construct the universal cover of the coset graph viewed as a cell complex.[9]

Triangulation. Upon modifying the coset graph if necessary by adjoining new vertices or otherwise, we might assume that the local cycles are triangles and, by Theorem 2.1, to prove the universality of a given partition one has to triangulate the cycles in the corresponding coset graph. This technique was pioneered by M. Ronan.[10]

Bounding the diameter. Here one starts with a vertex of the coset graphs and releases paths starting at this vertex. Then, by applying the equivalence with respect to changing direction of joining two vertices in a local cycle, we aim to show that every class of paths contains a 'short' one. This tool is essential for the Aschbacher–Segev 'uniqueness systems' approach. Notice that by means of this procedure one can establish the uniqueness of a group, but never its existence.

[9] S. Rees and L. H. Soicher, An algorithmic approach to fundamental group and covers of combinatorial cell complexes, *J. Symb. Comp.* **29** (2000), 59–77.

[10] M. A. Ronan, Covering and automorphisms of chamber systems, *Europ. J. Comb.* **1** (1980), 259–269.

Representation. In order to establish the existence of a faithful completion it is sufficient to embed the amalgam in any group, which might be large or even infinite. Thus one might try to construct an injection

$$\varphi \colon \mathcal{A} \to GL(\mathbb{C}) \cong GL_d(\mathbb{C})$$

for some vector space V of dimension d over the field $\mathbb{C}$ of complex numbers. To move towards this goal, we can start with a triple (χ_1, χ_2, χ_3) of $\mathbb{C}$-valued characters of A_1, A_2 and A_3, respectively, such that

$$\chi_i(g) = \chi_j(g) \text{ for } g \in A_i \cap A_j \text{ and } 1 \leq i < j \leq 3.$$

This enables us to construct representations of the subamalgam $\{A_i, A_j\}$ for any pair of indices, but there is no guarantee that we can represent all three members coherently. This is possible when an amalgam is constrained and Thompson's uniqueness criterion holds:

$$n_1 + n_2 = n_{12} - 1,$$

where n_1, n_2 and n_{12} are the dimensions of the centralizers in $GL(\mathbb{C})$ of $\varphi(A_1)$, $\varphi(A_2)$ and $\varphi(A_1 \cap A_2)$, respectively.[11] For the Mathieu amalgam introduced in the next section the uniqueness criterion is satisfied in the 45-dimensional irreducible module, which is essential for the construction accomplished in Chapter 7.

2.5 The Mathieu Amalgam

In this section we outline our construction procedure and many of the assertions made here will be proved later. We define the *Mathieu amalgam* to be a rank-3 amalgam $\mathcal{A} = \{G_1, G_2, G_3\}$ with the isomorphism types of its members being as described in the *parabolic structure* of Theorem 1.8, that is

$$G_1 \cong 2^4 : L_4(2), \quad G_2 \cong 2^6 : (L_3(2) \times S_3), \quad G_3 \cong 2^6 : 3 \cdot S_6$$

subject to the indices of intersections described by the Table 2.2, where the entry in the ith row and jth column is the index $[G_i : G_{ij}]$.

The Mathieu amalgam is constrained with respect to the largest solvable normal subgroup $T_3 \cong 2^6 : 3$ of G_3. The subgroup T_3 has trivial centre and hence, by Theorem 1.8, the isomorphism type of the Mathieu amalgam is uniquely determined by the isomorphism type of $\{G_1, G_2\}$ (subject to the conditions (C1) to (C4) with respect to T_3).

[11] J. G. Thompson, Finite dimensional representations of free products with an amalgamated subgroup, *J. Algebra* **69** (1981), 146–149.

Table 2.2

	G_{1i}	G_{2i}	G_{3i}
G_1	1	15	35
G_2	3	1	7
G_3	15	15	1

Our construction of the Mathieu groups is based on an independent self-contained proof of the fact that the Mathieu amalgam exists and is unique, and that its universal completion is the Mathieu group M_{24} as we knew it, that is a simple group of order $2^{10} \cdot 3^3 \cdot 5 \cdot 7 \cdot 11 \cdot 23$, which acts 5-fold transitively on a 24-set. Since the existence of M_{24} is not assumed *a priori*, this gives a new construction of M_{24} based on the amalgam method.

We construct the Mathieu amalgam by twisting the amalgam of three maximal parabolic subgroups in the classical group $H \cong GL_5(2)$ of invertible 5×5 matrices with entries in $GF(2)$. Let V be a five-dimensional $GF(2)$-space, viewed as a natural module for H, let

$$0 < V_1 < V_2 < V_3 < V_4 < V_5 = V$$

be a maximal flag in V, and let H_i be the stabilizer of V_i in H for $i = 1$, 2 and 3. Then

$$H_1 \cong 2^4 : L_4(2), \quad H_2 \cong 2^6 : (L_3(2) \times S_3), \quad H_3 \cong 2^6 : (L_3(2) \times S_3).$$

The subgroup H_{12} possesses an outer automorphism α, which enables us to construct the amalgam $\{G_1, G_2\}$ by applying the α-twist of $\{H_1, H_2\}$ in the sense of Goldschmidt's lemma. Thus $\{G_1, G_2\}$ and $\{H_1, H_2\}$ have the same type and G_1 can be identified with H_1. Let $R_3 \cong 2^6$ be the largest normal 2-subgroup in H_3. Because of the identification we have made, R_3 is a subgroup of $G_{12} = H_{12}$ and we can produce in $\{G_1, G_2\}$ a subamalgam

$$\mathcal{A}_3 = \{N_{G_1}(R_3), N_{G_2}(R_3)\}.$$

The core T_3 of this subamalgam is R_3 extended by an order-3 group acting on R_3 fixed-point freely. The centralizers of T_3 in every member of the subamalgam $\mathcal{A}_3$ are trivial and

$$\mathrm{Aut}(T_3) \cong 2^6 : \Gamma L_3(4).$$

Thus the Mathieu amalgam is constrained with respect to T_3 (whose centre is trivial) and therefore we define G_3 to be the subgroup in $\mathrm{Aut}(T_3)$ generated by the members of $\mathcal{A}_3$. The core T_3 is the kernel of A_3 acting on the projective plane Ξ of order 4, while $H_3 = N_H(T_3)$ is the stabilizer of a Fano sub-plane in this action. The α-twist transfers it to the stabilizer of a hyperoval, which forces G_3 to gain the required isomorphism type $2^6 : 3 \cdot S_6$. It is immediate from the definition of G_3 that it intersects G_1 and G_2 in exactly the members of the subamalgam $\mathcal{A}_3$. By comparing the orders we obtain the required diagram of indices. At this stage the construction of the Mathieu amalgam $\mathcal{A}$ will be complete.

According to the orthodox form of the amalgam method one should accomplish two formally independent tasks:

(i) show that $\mathcal{A}$ possesses at least one faithful completion;
(ii) show that the index of G_1 in the universal completion is 759.

In the situation being considered here it proved convenient to accomplish these two steps coherently. We show that $\mathcal{A}$ possesses a faithful permutation representation of degree 24 in the sense that there is a faithful completion

$$\psi : \mathcal{A} \to S_{24}.$$

Next, by comparing the G_1-parts of the coset graphs of the universal completion G of $\mathcal{A}$ and the completion ψ in S_{24}, we demonstrate that they are isomorphic, supporting the structure of a distance-transitive graph with the intersection diagram of the octad graph. This shows that ψ is universal and that G is M_{24} (as we knew it to be).

3

$L_4(2)$ in Two Incarnations and $L_3(4)$

There are two finite simple groups of order 20 160 and both are cornerstones in the theory of sporadic simple groups. One of them appears in two incarnations: as the alternating group A_8 of degree eight and as the projective special linear group $L_4(2)$ in dimension 4 over the field of two elements. The other one is $L_3(4)$ (sometimes denoted by M_{21} to indicate its importance for the Mathieu groups). These two groups and their subgroups are discussed in this chapter. First we illustrate the amalgam method for proving the isomorphism $A_8 \cong L_4(2)$ by recovering the 2-local parabolic subgroups given the alternating incarnation of the group. The central point of the chapter is the proof of the *double complement* Lemma 3.10 for $2^3 : L_3(2)$ viewed as a parabolic subgroup in $L_4(2)$. We prove the existence and produce an explicit form of the outer automorphism of $2^3 : L_3(2)$, which permutes the classes of $L_3(2)$-complements. In a certain sense the group $L_3(2)$, which also has an $L_2(7)$ incarnation, is the main character of this book's plot.

3.1 The Fano Plane

Let $V = V_3(2)$ be the three-dimensional $GF(2)$-space. We call the seven non-zero vectors of V *points* and denote the point set by P. Every two-dimensional subspace of V contains three points; we call such triples *lines* and denote the line set by L. A triple of points is a line whenever the sum of the points is the zero vector.

Since a non-zero vector of a $GF(2)$-space can be identified with the one-dimensional subspace it spans, $\Phi = (P, L)$ is the projective plane associated with V, also known as the *Fano plane*, which is described by the famous diagram shown in Figure 3.1.

The following properties of Φ implied by the rank and nullity principle will be axiomatized:

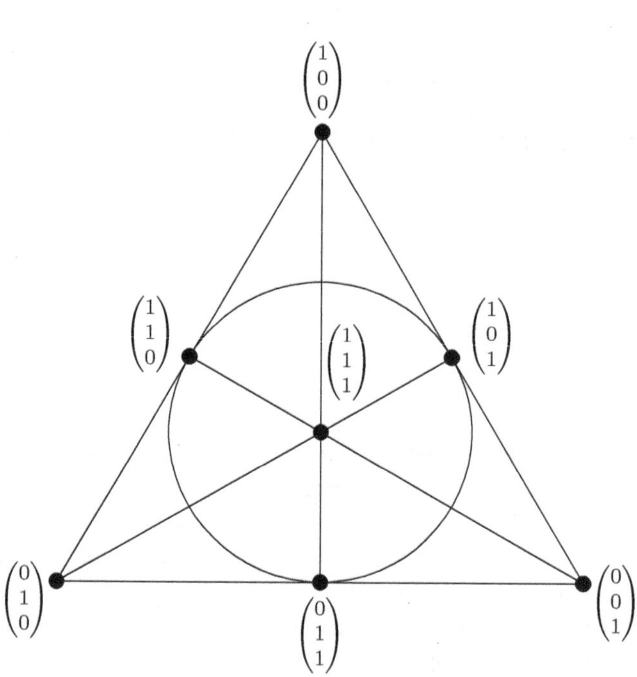

Figure 3.1

(a) P is a set of seven elements called *points*;
(b) L is a collection of seven three-element subsets of P called *lines*;
(c) every point is on exactly three lines;
(d) any two distinct points are on a unique line;
(e) any two distinct lines have a unique point in common.

Subject to these axioms $\Phi = (P, L)$ is said to be a *projective plane of order* 2.

Lemma 3.1 *The Fano plane represents the unique isomorphism class of projective planes of order* 2.

Proof By (a) we start with a seven-element set P of points and show that, up to renaming, the collection L of lines is determined uniquely. Take a 3-subset l in P and declare it to be one of the lines. Then by (b) $D := P \setminus l$ consists of four points, by (e) the set D does not contain lines, by (d) each of the six pairs in D determines a unique line, and by (b) these six lines together with l comprise the line-set of Φ. Whenever two pairs are disjoint, they form a partition of D, and their intersection is a point on l. Since the six pairs in D fall into three partitions, in order to complete Φ it remains to match these partitions to the three points on l, which can be done uniquely, up to renaming. □

When we consider L as an abstract seven-element set and identify a point from P with the triple of lines through this point, we obtain the plane $\Phi^* = (L, P)$, which is dual to Φ. It is immediately evident that Φ^* is also a projective plane of order 2, and hence isomorphic to Φ by Lemma 3.1.

3.2 The Group $L_3(2)$

Let $V = V_3(2)$ be a three-dimensional $GF(2)$-space as in the previous section and let

$$F := GL(V) \cong GL_3(2) \cong L_3(2)$$

be the general linear group of V. Then F is the automorphism group of the additive group of V, which is an elementary abelian group of order eight. Being the general linear group of V, the group F acts regularly on the set of bases of V, so that

$$|F| = (2^3 - 1)(2^3 - 2)(2^3 - 2^2) = 2^3 \cdot 3 \cdot 7 = 168.$$

On the other hand, the natural action of F on P preserves the linear dependence relation, and hence preserves L as a set, so that $F \leq \mathrm{Aut}(\Phi)$. Actually F must be the full automorphism group of Φ, since V can be uniquely recovered from Φ in the following way: the vector-set of V is P together with the zero vector and the addition rule is dictated by the lines (the sum of two points on a line is the third one). Thus

$$F = \mathrm{Aut}(\Phi).$$

As F is a linear group, it is immediately evident that its action on the line-set L is transitive, since these are just the two-dimensional subspaces of V, and we take a closer look at the stabilizer $F(l)$ of a line $l \in L$. The uniqueness proof for Φ in the previous section involves an encoding of Φ by a pair $(l \cup D, \varphi)$, consisting of a partition of P into a line l and its complement D together with a matching φ which assigns to every partition of D into disjoint pairs a point on l. It is easy to see that $F(l)$ stabilizes this encoding, acts faithfully on D, and realizes every permutation of D. Thus $F(l) \cong \mathrm{Sym}(D) \cong S_4$, which is consistent with the following numerology:

$$|F| = |L| \cdot |F(l)| = 7 \cdot 4! = 168.$$

The action of $\mathrm{Sym}(D)$ on the six 2-subsets in D is transitive, while these subsets are in bijection with the lines in L other than l. Therefore F acts on L

doubly transitively as on the cosets of S_4. Because of the isomorphism between the plane $\Phi = (P, L)$, and its dual $\Phi^* = (L, P)$, we have the dual version of this setting. The dual version of the statement about the structure of the point stabilizer $F(p)$ for $p \in P$ is clear from the linear setting and the isomorphism

$$S_4 \cong AGL_2(2).$$

Thus, for a point $p \in P$ and a line $l \in L$, we obtain two non-conjugate maximal subgroups in F, each being isomorphic to S_4. Consider their possible intersections. If $p \in l$ then $F(p) \cap F(l) \cong D_8$ is the stabilizer of the maximal flag

$$0 < p < l < V,$$

which is known as the *Borel subgroup* and is a Sylow 2-subgroup in F. If $p \notin l$ then $F(p) \cap F(l) \cong S_3 \cong L_2(2)$ is the stabilizer of the direct sum decomposition

$$V = p \oplus l$$

known as a *Levi complement* in $F(p)$ and $F(l)$ to their *radicals*, which are the Klein 4-subgroups of the S_4s.

By the above consideration F, acting on the point–line pairs (p, l), has two orbits: the orbit of incident pairs called *flags* of length 21 with stabilizer D_8 and the orbit of non-incident pairs called *anti-flags* of length 28 with stabilizer $S_3 \cong D_6$.

If V^* is the dual space of V (i.e. the space of linear functions on V) then Φ^* is the projective plane associated with V^*. Since F is also the general linear group of V^*, an isomorphism between Φ and Φ^* can be realized as a group automorphism of F. In terms of matrices, V can be realized by column vectors on which F acts from the left and V^* by row vectors on which F acts from the right. Then a duality automorphism is performed by transposition of a matrix followed by its inversion. In this way we obtain the *inverse–transpose automorphism*, which is an outer automorphism of F.

3.3 Elements and Subgroups of $L_3(2)$

We are aiming to classify the elements of F up to conjugacy. We start with *involutions* (elements of order 2). Since F acts transitively on the set L and hence on the set

$$\{F(l) \mid l \in L\}$$

of line stabilizers via conjugation, and $|L| = 7$ is an odd number, every involution in F is conjugate to an involution τ in $F(l) \cong S_4$. Having chosen $l \in L$ we determine Φ by the encoding $(l \cup D, \varphi)$ associated with l. There are two classes of involutions in $\mathrm{Sym}(D)$: six transpositions and three non-identity elements of the Klein 4-subgroup. If τ is one of the latter, then its orbits on D form a partition of D into two pairs and the point p on l matched to this partition via φ is fixed. Then τ is the *transvection* $\tau(p,l)$ with centre p and axis l, which, by definition, is the automorphism which fixes l point-wise and adds p to every point outside l. Suppose now that τ acts on D as a transposition. Then τ still acts on the whole of P as a transvection whose axis is the line m through the pair of points in D fixed by τ and whose centre is the intersection of m with l.

Thus F contains a single class of involutions of size 21 consisting of the transvections indexed by the flags of Φ with centralizers isomorphic to D_8. Already inside $F(l) \cong S_4$ we can find pairs of involutions with products of order 1, 2, 3 and 4. Arguing as in the case of involutions, we observe that an element of order 4 is conjugate to an element δ in $F(l)$ and l is the only line stabilized by δ. This gives the unique class of 42 elements of order 4.

Since $|P| = |L| = 7$, which is 1 modulo 3, we conclude that every element σ of order 3 in F stabilizes a point p and a line l. By the last paragraph of the previous section (p,l) must be an anti-flag. Thus F contains a single class of order-3 elements of size 56 and the subgroups of order 3 are indexed by the anti-flags.

Now it remains only to analyze an element ε, whose order is divisible by 7.

Since 7 does not divide the order of the centralizer of an element of order 2 or 3, the order of ε must be precisely 7, and such an element exists by Cauchy's theorem since 7 is a prime number dividing the order of F. The subgroup generated by ε is a Sylow 7-subgroup of F, and its normalizer S has index in F congruent to 1 modulo 7, which must be 8. Now it is clear that F conjugates the Sylow 7-subgroups in a doubly transitive manner, so that S is another maximal subgroup in F, isomorphic to the Frobenius group of order 21, and there are two classes of order-7 elements in F of size 24, each with an element and its inverse belonging to different classes. Thus we have the following.

Lemma 3.2 *The group F contains six conjugacy classes of elements with representatives 1, τ, σ, δ, ε and ε^{-1}, whose sizes are the summands on the right-hand side of the equality*

$$168 = 1 + 21 + 56 + 42 + 24 + 24,$$

and by Lagrange's theorem the above equality easily shows that F is a simple group. □

Lemma 3.3 *The group F contains three classes of maximal subgroups with representatives*

$$F(p) \cong S_4, \quad F(l) \cong S_4 \ \text{and} \ S = N_F(\langle \varepsilon \rangle) \cong F_{21}$$

and indices 7, 7 *and* 8, *respectively.*

Proof Let X be a maximal subgroup in F. Suppose first that the order of X is divisible by 7. If X contains a single Sylow 7-subgroup, then it is contained in a conjugate of S, otherwise it contains eight Sylow 7-subgroups, which generate F, since the latter is simple. Thus we may assume that the only primes which might divide the order of X are 2 and 3. Therefore the action of X on P is intransitive, so that its shortest orbit O has length at most 3. If the length is 1, then X stabilizes a point; if the length is 2, then X stabilizes the line determined by this orbit; if the length is 3, then either it is a line, or the sum over the orbit is a non-zero point stabilised by X. In any event X is contained either in a conjugate of $F(p)$ or in a conjugate of $F(l)$. □

Let us now demonstrate how Φ can be recovered from the abstract structure of the group F. Since F possesses the inverse–transpose automorphism, the best we can hope to do is to reconstruct Φ up to duality, and this can be achieved in the following way. Let τ_1 and τ_2 be involutions in F. Whether or not τ_1 and τ_2 commute is an abstract feature. On the other hand, we know that these involutions are transvections of Φ, so that

$$\tau_1 = \tau(p_1, l_1), \quad \tau_2 = \tau(p_2, l_2).$$

It can easily be checked that $[\tau_1, \tau_2] = 1$ if and only if $p_1 = p_2$ or $l_1 = l_2$ or both (in the latter case $\tau_1 = \tau_2$). Thus a maximal set X of pairwise commuting involutions is of size 3, there are 14 such sets and they are in an F-correspondence with $P \cup L$. The set X can be split into two equal pieces either by considering the F-orbits on X or by defining a graph on X where two subsets are adjacent whenever they are disjoint: the required splitting is the unique bipartition of the graph. Now, by declaring which of the two F-invariant 7-subsets of X are points and which are lines and by defining two objects of different type incident if they intersect properly (as triples of involutions), we obtain either Φ or Φ^*. Besides other important aspects, this shows that the outer automorphism group of F has order 2 and thus every outer automorphism is the product of the inverse–transpose automorphism with an inner automorphism.

3.4 An Automorphism of $2^3 : L_3(2)$

With $V = V_3(2)$ being a three-dimensional $GF(2)$-space, and $F = GL(V) \cong L_3(2)$ being the general linear group of V, let M be the group of affine transformations of V. Then

$$M = \{(f, v) \mid f \in F, v \in V\},$$

and the action of M on V is described by the following rule:

$$(f, v)\colon u \mapsto u^f + v,$$

so that

$$(f, v) \cdot_M (g, u) = (f \cdot_F g, v^g +_V u).$$

Thus M is the semidirect product of V and F (commonly denoted by $2^3 : L_3(2)$) with respect to the natural action.

A crucial role in our further exposition and in the whole of finite group theory is played by the fact that M contains two conjugacy classes of complements to V, and therefore possesses an outer automorphism permuting these two classes. One of the ways of accomplishing this task is to show that M contains the projective special linear group $L_2(7)$ over the field $GF(7)$. This is the subgroup which constitutes a 'non-standard' complement (which is not conjugate to F in M). Our proof of the isomorphism $L_3(2) \cong L_2(7)$ is constructed along the same lines as in Conway's brilliant article.[1]

Let us specify the standard setting for the natural action of $L_2(7)$ on the projective $GF(7)$-line $\mathcal{J}$. The projective line $\mathcal{J}$ is the set of eight one-dimensional subspaces of the two-dimensional $GF(7)$-space (with a standard basis). In each such subspace we choose a 'canonical' vector by applying the rule that the leading non-zero entry is one. This gives

$$\mathcal{J} = \{\infty\} \cup GF(7),$$

where ∞ stands for the subspace with canonical vector $\binom{0}{1}$, and $\alpha \in GF(7)$ for the one with canonical vector $\binom{1}{\alpha}$. The group $L_2(7)$ is the image of the action on $\mathcal{J}$ of the group $SL_2(7)$ of 2×2 determinant-1 matrices with $GF(7)$-entries. One can also realize the elements of $\mathcal{J}$ as tangents of the respective lines in the standard basis. The stabilizer of ∞ is the image U of the group

[1] J. H. Conway, Three lectures on exceptional groups, in *Finite Simple Groups*, ed. M. B. Powell and G. Higman, Academic Press, New York, 1971, pp. 215–247.

of lower triangular matrices, and in order to generate the whole of $L_2(7)$ it suffices to adjoin to U the image σ of the matrix

$$\begin{pmatrix} 0 & 1 \\ -1 & 0 \end{pmatrix},$$

which permutes ∞ with 0 and sends $\alpha \in GF(7) \setminus \{0\}$ onto $-\alpha^{-1}$, so that

$$\sigma = (\infty\, 0)(1\, 6)(2\, 3)(4\, 5).$$

The group U is the Frobenius group of order 21 with kernel and complement generated by the images of the matrices

$$t(\alpha) = \begin{pmatrix} 1 & 0 \\ \alpha & 1 \end{pmatrix} \quad \text{and} \quad s(\beta) = \begin{pmatrix} \beta & 0 \\ 0 & \beta^{-1} \end{pmatrix},$$

respectively, where $\alpha, \beta \in GF(7)$ and $\beta \neq 0$. Both $t(\alpha)$ and $s(\beta)$ stabilize ∞; $t(\alpha)$ adds α and $s(\beta)$ multiplies by β^2 the remaining elements of $\mathfrak{J}$.

Let us now view the Fano plane as a quadratic residue design by putting

$$P = GF(7) = \mathfrak{J} \setminus \{\infty\}, \quad L = \{\{1+q, 2+q, 4+q\} \mid q \in GF(7)\},$$

where one immediately recognizes in $Q := \{1, 2, 4\}$ the set of non-zero squares in $GF(7)$. Thus defined, the design (P, L) is manifestly U-invariant. For instance

$$(Q+q)^{s(2)} = Q^{t(q)s(2)} = Q^{s(2)t(2q)} = Q^{t(2q)} = Q + 2q.$$

The pair $\Phi = (P, L)$ is indeed a Fano plane, since it can be encoded as $(Q \cup (P \setminus Q), \varphi)$, where

$$\varphi\colon \{0,3\} \cup \{5,6\} \mapsto 1; \quad \varphi\colon \{0,6\} \cup \{3,5\} \mapsto 2; \quad \varphi\colon \{0,5\} \cup \{3,6\} \mapsto 4.$$

Thus there is a bijective mapping ψ of $\mathfrak{J} \setminus \{\infty\}$ onto the set of non-zero vectors of a three-dimensional $GF(2)$-space V which commutes with the action of U. We extend ψ to a U-invariant bijection of $\mathfrak{J}$ onto V by defining $\psi(\infty)$ to be the zero vector. Now, in order to obtain the desired inclusion $L_2(7) \leq M$, it is sufficient to show that M contains σ. The element σ permutes ∞ and 0. Since $\{0, 1, 3\}$, $\{0, 2, 6\}$ and $\{0, 5, 4\}$ are the lines of Φ containing the zero element of $GF(7)$, the permutation

$$\pi = (\infty\, 0)(1\, 3)(2\, 6)(4\, 5)$$

is contained in M, since it is the translation by that element. Finally, the product $\tau = (1\, 2)(3\, 6)$ of π by σ is readily seen to be the transvection of Φ with axis Q and centre $\{4\}$, thus it is contained in M and hence σ is also in M.

Lemma 3.4 *The group $M \cong 2^3 : L_3(2)$ of affine transformations of $V = V_3(2)$ contains a subgroup which is isomorphic to $L_2(7)$ and acts doubly transitively on the set of eight vectors in V.* $\square$

3.5 Another Look at the Isomorphism $L_3(2) \cong L_2(7)$

As in Section 3.4, let $\mathfrak{J} = \{\infty\} \cup GF(7)$ be the projective line over $GF(7)$. Then $L_2(7)$ acts naturally on $\mathfrak{J}$ and we discuss the structure of the corresponding $GF(2)$-permutation module $2^{\mathfrak{J}}$. This discussion will again demonstrate the isomorphism of $L_2(7)$ with $L_3(2)$ and provide some useful information on some indecomposable extensions.

Let $Q = \{1, 2, 4\}$ and $N = \{3, 5, 6\}$ be the set of non-zero squares and non-squares in $GF(7)$. Let I_Q and I_N be the smallest submodules in $2^{\mathfrak{J}}$ containing $\{\infty\} \cup Q$ and $\{\infty\} \cup N$, respectively.

Lemma 3.5 *The eight-dimensional $GF(2)$-permutation module $2^{\mathfrak{J}}$ of $L_2(7)$ acting on the projective $GF(7)$-line $\mathfrak{J}$ possesses just two composition series:*

$$0 < \{\emptyset, \mathfrak{J}\} < I_A < X^{\perp} < 2^{\mathfrak{J}},$$

where A is $Q = \{1, 2, 4\}$ or $N = \{3, 5, 6\}$, and I_A/X is an indecomposable extension of the trivial one-dimensional module X by a three-dimensional module.

Proof Let X be a minimal non-zero $L_2(7)$-submodule in $2^{\mathfrak{J}}$. Since $|X \setminus \{0\}|$ is odd, a Sylow 2-subgroup S in $L_7(2)$ fixes a non-zero vector in X. On the other hand, $|S| = |\mathfrak{J}| = 8$, while the element stabilizer in $L_2(7)$ (which is the Frobenius group F_{21} of order 21) intersects S trivially. Thus the whole of $\mathfrak{J}$ is the only non-zero vector fixed by S and $X = \{\emptyset, \mathfrak{J}\}$ is the trivial submodule formed by the two improper subsets of $\mathfrak{J}$. Clearly $X^{\perp}$ is the submodule of even subsets in $\mathfrak{J}$, which contains X, and we claim that the factor-module $X^{\perp}/X$ is the direct sum of two three-dimensional $L_2(7)$-submodules. Indeed, one can immediately check that both I_Q and I_N are four-dimensionaly intersecting on $X = \{\emptyset, \mathfrak{J}\}$. $\square$

Since $L_2(7)$ has been shown to possess a three-dimensional faithful module, the isomorphism $L_2(7) \cong L_3(2)$ follows from the fact that $|L_2(7)| = |L_3(2)|$. It is noticeable that there is no uniserial five-dimensional $GF(2)$-module Z for $L_3(2)$ with a one-dimensional submodule X and a submodule Y of co-dimension 1, such that Y/X is irreducible (and thus is the natural or the dual natural module). In fact, by the *double complement* Lemma 3.10 Z/X

is the unique indecomposable extension of Y/X by the one-dimensional trivial module, and hence $Z/X \setminus Y/X$ is an eight-element set possessing an $L_3(2)$-invariant identification with $\mathcal{J}$. Since the stabilizer in $L_3(2)$ of an element from $\mathcal{J}$ is the Frobenius group F_{21}, which contains no index-2 subgroups, $Z \setminus Y$ is the union of two copies of $\mathcal{J}$. Each copy must generate the whole of Z because Z is uniserial, which forces Z to be a submodule of $2^{\mathcal{J}}$. By virtue of the structure of the permutation module there are no such submodules.

3.6 A_7 and a Space of Fano Planes

In this section we demonstrate that the set of Fano plane structures on a seven-element set carries the structure of a four-dimensional $GF(2)$ space. In order to accomplish this task we will make use of the following well-known characterization of projective $GF(2)$ geometries.

Proposition 3.6 *Let $\mathcal{P}$ be a finite set, let $\mathcal{L}$ be a collection of three-element subsets of $\mathcal{P}$, and let Π be a collection of seven-element subsets of $\mathcal{P}$ such that the following statements hold:*

(i) *any two elements of $\mathcal{P}$ are in a unique triple from $\mathcal{L}$;*
(ii) *for every 7-tuple $\pi \in \Pi$ the structure $(\pi, \mathcal{L}(\pi))$ is a Fano plane, where $\mathcal{L}(\pi)$ is the set of triples from $\mathcal{L}$ contained in π;*
(iii) *every triple of elements of $\mathcal{P}$ not contained in $\mathcal{L}$ is contained in a unique 7-tuple from Π.*

Then $\mathcal{P}$, $\mathcal{L}$ and Π are one-, two- and three-dimensional subspaces of a vector space V over $GF(2)$ with

$$\dim(V) = \log_2(|\mathcal{P}| + 1).$$

Proof We define the vector-set of V to be $\mathcal{P}$ together with the zero vector. Then $\mathcal{L}$ determines the addition rule: the sum of two distinct points determines a unique line and their sum is the third point on that line. Finally the planes in Π ensure the associativity of the addition rule. □

Let us treat the point-set P of the Fano plane Φ as an abstract seven-element set and consider the set $\mathcal{F}$ of all the Fano plane structures having P as the point-set. Since all Fano planes are isomorphic, the symmetric group $\mathrm{Sym}(P) \cong S_7$ acts transitively on $\mathcal{F}$ and the stabilizer of Φ in this action is $F = \mathrm{Aut}(\Phi) \cong L_3(2)$. By the orbit-stabilizer lemma

$$|\mathcal{F}| = |S_7|/|L_3(2)| = 30.$$

It follows from the simplicity of F that it is contained in the alternating subgroup $\mathrm{Alt}(P) \cong A_7$ and hence A_7 has two orbits on $\mathcal{F}$, which we denote by $\mathcal{P}$ and Π, of size 15 each. Let $\mathcal{L}$ be the set of partitions of P into subsets of size 3 and 4. Then $\mathcal{L}$ is comprised of 35 partitions, and $\mathrm{Alt}(P)$ acts transitively on $\mathcal{L}$ with stabilizer isomorphic to

$$(S_3 \times S_4)^+.$$

Here and elsewhere the superscript $+$ indicates the intersection with the alternating group, which is clear from the context.

In order to recover a Fano plane structure from a partition $P = l \cup D$, which is an element of $\mathcal{L}$, one needs to specify the matching φ, and this can be done in $6 = 3!$ different ways. These matchings can be obtained from a particular one by applying permutations from the symmetric group of l. In this way it clear that half of the elements lead to points in $\mathcal{P}$ and the other half to planes in Π.

Let us say that a partition $P = l \cup D$ from $\mathcal{L}$ is incident to a plane Ψ if l is a line of Ψ. Next we define an incidence relation on the set of planes. Let $p \in P$ and let l_1, l_2 and l_3 be the lines of Φ passing through p. Then each of the four lines of Φ will satisfy the following *intersection property*.

Lemma 3.7 *Every line of Φ is a three-element set of points which intersects each l_i for $1 \leq i \leq 3$ in a single point.* □

Let Y be the set of all triples satisfying the intersection property, and let Q be the largest subgroup of $\mathrm{Sym}(P)$ which stabilizes set-wise each l_i. Then both Y and Q have size 8 and Q acts regularly on Y. On the other hand, the intersection of Q with $\mathrm{Alt}(P)$ is precisely the Klein 4-subgroup of $F(p) \cong S_4$, and under its action Y splits into two orbits of length 4 each. One of the orbits is precisely the set of lines of Φ disjoint from p; the other orbit, together with the l_is, forms the line-set of a Fano plane Ψ which we declare to be incident to Φ. The incident planes Φ and Ψ are contained in different $\mathrm{Alt}(P)$-orbits and we extend the incidence relation via the action of $\mathrm{Alt}(P)$. Notice that, whenever two planes are incident to a common partition from $\mathcal{L}$, they are incident to each other, and that two incident planes are incident to exactly three common partitions from $\mathcal{L}$.

Before proceeding further we make one more observation. Let $(l \cup D, \varphi)$ be an encoding of Φ which makes it incident to the partition $P = l \cup D$ from $\mathcal{L}$. If Ψ is another Fano plane from the same $\mathrm{Alt}(P)$-orbit also incident to $P = l \cup D$, then Ψ is encoded by $(l \cup D, \varphi\pi)$, where π is a 3-cycle on l. This implies the observation we are after: Φ and Ψ share l but no further lines, in particular $P = l \cup D$ is the only partition in $\mathcal{L}$ incident to both Φ and Ψ. An easy counting procedure shows that Φ is contained in 14 triples

$$(\Phi, P = l \cup D, \Psi)$$

as above. Since 14 is the number of planes in the $\mathrm{Alt}(P)$-orbit of Φ other than Φ, we conclude that any two planes from the same $\mathrm{Alt}(P)$-orbit are incident to a common partition from $\mathcal{L}$.

Now, by Proposition 3.6, we have the following lemma.

Lemma 3.8 *The above-constructed triple $(\mathcal{P}, \mathcal{L}, \Pi)$ is the projective geometry of a four-dimensional $GF(2)$-space on which $\mathrm{Alt}(P) \cong A_7$ acts flag-transitively and elements from $\mathrm{Sym}(P) \setminus \mathrm{Alt}(P)$ perform duality automorphisms.* □

3.7 $L_4(2) \cong A_8$

Let $V_4(2)$ be a four-dimensional $GF(2)$-space. It is not a secret that the order of $GL(V_4(2)) \cong L_4(2)$ is half of $8!$, which is the order of A_8. Since $L_4(2)$ is the automorphism group of the projective geometry of $V_4(2)$, by Lemma 3.8 $L_4(2)$ contains a subgroup isomorphic to A_7 and, since

$$[L_4(2) \colon A_7] = [A_8 \colon A_7] = 8,$$

we can define a homomorphism of $L_4(2)$ into $\mathrm{Sym}(\Omega)$, where Ω is the set of A_7-cosets in $L_4(2)$. The image of the homomorphism has index 2 in $\mathrm{Sym}(\Omega)$; it must be $\mathrm{Alt}(\Omega)$, and this immediately gives the famous isomorphism

$$L_4(2) \cong A_8.$$

In the remaining part of the section this isomorphism is brought to an explicit form by identifying the elements of the projective geometry of $V_4(2)$ with maximal elementary abelian subgroups of A_8.

Let Ω be an abstract eight-element set, and let $S(\Omega)$ and $A(\Omega)$ denote the symmetric and alternating groups of Ω. Let E be a regular elementary abelian subgroup (of order 8) in $S(\Omega)$. It is clear that all such subgroups are conjugate in $S(\Omega)$, each being self-centralized and fully normalized in the sense that

$$C_{S(\Omega)}(E) = E \quad \text{and} \quad N_{S(\Omega)}(E)/C_{S(\Omega)}(E) \cong \mathrm{Aut}(E) \cong L_3(2).$$

It is also clear that the stabilizer K in $N_{S(\Omega}(E)$ of an element from Ω complements E in the normalizer and is isomorphic to $L_3(2)$. Because of the simplicity of $L_3(2)$ and the irreducibility of $GL(E) \cong L_3(2)$ on E, $N_{S(\Omega)}(E)$ is contained in $A(\Omega)$. Thus $S(\Omega)$ contains a single class of

$$30 = [S(\Omega) : N_{S(\Omega)}(E)]$$

regular elementary abelian subgroups of order 8, and this orbit splits under $A(\Omega)$ into two orbits of size 15 each. The latter two orbits (denoted by $\mathcal{P}$ and Π by analogy with the previous section) are going to be the sets of one- and three-dimensional subspaces in the space $V_4(2)$ to be associated with $A(\Omega)$. The two-dimensional subspaces are formed by the set $\mathcal{L}$ of partitions of Ω into two subsets of size 4 each. There are 35 such partitions, which is exactly the number of 2-subspaces in $V_4(2)$.

We are about to demonstrate a modified version of the isomorphism proof between $A(\Omega)$ and $L_4(2)$ and redefine the incidence relation on

$$\mathcal{P} \cup \mathcal{L} \cup \Pi$$

in terms of stabilizers in $A(\Omega)$ of the corresponding elements.

If E_1 is a regular elementary abelian subgroup of $A(\Omega)$, which is an element of $\mathcal{P}$, then its stabilizer M_1 in $A(\Omega)$ is the normalizer $N_{A(\Omega)}(E_1)$ and the latter is a semidirect product of E_1 and a complement $K_1 \cong L_3(2)$, with K_1 being the stabilizer in M_1 of an element of Ω. The subgroup E_1 will be called the *radical* of M_1 and in view of the $S(\Omega)$-symmetry we have

$$M_1 = E_1 : K_1 \cong 2^3 : L_3(2) \cong E_3 : K_3 = M_3.$$

There are 105 fixed-point free involutions in $A(\Omega)$, and each of the 30 subgroups in $\mathcal{P} \cup \Pi$ (which are now called radicals of order 8) contains seven of them. Hence every fixed-point free involution is in precisely two such radicals, and by the symmetry between $\mathcal{P}$ and Π these two radicals are in different $A(\Omega)$-orbits (one in $\mathcal{P}$ and one in Π). We declare these two radicals to be incident. This can be restated in the following form: $E_1 \in \mathcal{P}$ and $E_3 \in \Pi$ are incident if and only if the following three equivalent conditions hold:

$$|E_1 \cap E_3| = 2, \quad E_3 \leq M_1, \quad E_1 \leq M_3.$$

Let us now analyse the structure of the stabilizer M_2 of a partition $\Omega = \Delta_1 \cup \Delta_2$ which is an element of $\mathcal{G}_2$ (so that $|\Delta_1| = |\Delta_2| = 4$). If E_2 is the direct product of the Klein 4-subgroups of $\mathrm{Sym}(\Delta_1)$ and $\mathrm{Sym}(\Delta_2)$ then E_2 is elementary abelian of order 2^4 normal in M_2 and $M_2 = N_{A(\Omega)}(E_2)$. If T_2 is a Sylow 3-subgroup of M_2 then T_2 is elementary abelian of order 9 generated by a 3-cycle a on Δ_1 and a 3-cycle b on Δ_2, so that T_2 acts on E_2 fixed-point freely. Next, a Sylow 2-subgroup of $N_{M_2}(T_2)$ is elementary abelian of order 4 generated by the involutions

$$\sigma_1 : a \mapsto b \text{ and } \sigma_2 : a \mapsto b^{-1}.$$

Then $K_2 := N_{M_2}(T_2)$ is the direct product of two subgroups $\langle ab^{-1}, \sigma_1\rangle$ and $\langle ab, \sigma_2\rangle$, each isomorphic to S_3, and the whole of M_2 is the semidirect product of E_2 and K_2. If each of the direct factors of K_2 is considered as an $L_2(2)$-subgroup, then E_2 bears the tensor product structure of the natural modules of these factors. In short

$$M_2 = N_{A(\Omega)}(E_2) = E_2 : K_2 \cong 2^4 : (S_3 \times S_3).$$

In particular, $\mathcal{G}_2$ can be viewed as the set of $A(\Omega)$-conjugates of the radical E_2 of order 16.

We define the incidence between a radical $E \in \mathcal{P} \cup \Pi$ of order 8 and a radical E_2 of order 16 similarly to the previous case by the following condition:

$$|E \cap E_2| = 4.$$

Let us locate the radicals of order 8 inside M_2. The subgroup E_2 itself cannot contain a radical of order 8, since, unlike E_2, the radical E must act transitively on Ω. On the other hand, the whole of M_2 does contain radicals of order 8, since M_2 contains a Sylow 2-subgroup of $A(\Omega)$. Let σ be a fixed-point free involution in $M_2 \setminus E_2$, so that σ permutes Δ_1 and Δ_2. Then σ normalizes a Sylow 3-subgroup of M_2. In fact, to find such a subgroup we take a 3-cycle a on Δ_1, define $b = a^\sigma$ and observe that σ normalizes the subgroup T_2 of order 9 generated by ab and ab^{-1}. Furthermore, σE_2 is one of the six involutions contained in the direct factors of $M_2/E_2 \cong S_3 \times S_3$. Next $C_{E_2}(\sigma)$ is of order 4 with non-identity elements $k^\sigma k$, where k is an involution of the Klein 4-subgroup on Δ_1. It is easy to see that $\langle \sigma, C_{E_2}(\sigma)\rangle$ is indeed a radical of order 8. The outcome can be summarized as follows.

Lemma 3.9 *The following assertions hold:*

(i) *a radical of order* 16 *is incident to six radicals of order* 8, *of which three are in* $\mathcal{P}$ *and three are in* Π*;*
(ii) *two radicals of order* 8 *incident to a common radical of order* 16 *are contained in different* $A(\Omega)$*-orbits and incident to each other;*
(iii) *for a given radical* E *of order* 8 *the radicals of orders* 8 *and* 16 *incident to* E *are in a natural bijection with subgroups of orders* 2 *and* 4 *in* E. □

By Lemma 3.9 the conditions (i) to (iii) in Proposition 3.6 hold for the incidence relation defined on $\mathcal{P} \cup \mathcal{L} \cup \Pi$, which re-establishes the isomorphism

$$A(\Omega) \cong A_8 \cong L_4(2).$$

3.8 Subgroup Correspondence

Let $A \cong A_8 \cong L_4(2)$, let Ω be a set of size 8 on which A acts as the alternating group, so that $A = A(\Omega)$, and let $V = V_4(2)$ be a four-dimensional $GF(2)$-space on which A acts as the general linear group, so that $A = GL(V)$. In this section we discuss some important subgroups in A and describe their actions on Ω and on V.

The symmetric group $S(\Omega) \cong S_8$ is the automorphism group of the alternating group $A(\Omega) \cong A_8$, and hence it must be the extension of $L_4(2)$ by $\langle \sigma \rangle$, where σ is an isomorphism from V to V^*. As usual there is no canonical choice for such an isomorphism, since we may take σ to be any odd element of $S(\Omega)$. The inverse–transpose automorphism is realized by a transposition.

3.8.1 $L_3(2) \cong L_2(7)$

We start by classifying the subgroups in A which are isomorphic to $L_3(2)$, and as a by-product this will provide us with a classification of the four-dimensional $GF(2)$-representations of $L_3(2)$. Let K be an $L_3(2)$-subgroup in A. Consider the action of K on Ω. Every orbit of this action is equivalent to an action on the cosets of a subgroup, whose index is the length of the orbit. In particular, the index of such a subgroup is at most 8. The description of maximal subgroups in K given in Lemma 3.3 shows that either

(1) K fixes an element of Ω and acts on the remaining seven elements as on the cosets of an S_4-subgroup, or
(2) K acts transitively on Ω as on the set of its Sylow 7-subgroups via conjugation.

We claim that the subgroups within the sets (1) and (2) are conjugate in $S(\Omega)$. For (2) this is obvious; for (1) one might have a moment of weakness, thinking about two classes of S_4-subgroups in $L_3(2)$. But the situation should be clarified by the remark that a subgroup in (1) is the stabilizer in $S(\Omega)$ of an element of Ω together with a projective plane structure on the remaining elements. On the other hand, the subgroups in (1) are self-normalized in A, and hence inside A they split into two classes.

Table 3.1

	Ω	V	V^*
K_1	$1+7$	$V_1 \setminus V_3$	$V_3^* \setminus V_1^*$
K_2	8	$V_1 \oplus V_3$	$V_1^* \oplus V_3^*$
K_3	$1+7$	$V_3 \setminus V_1$	$V_1^* \setminus V_3^*$

From the discussion in the previous section it must be clear that the non-Levi complements K_1 and K_3 are representatives of these two classes. The relationship between the radicals of order 8 from different A-orbits demonstrates that K_1 does not stabilize a 3-subspace in V, while K_3 does not stabilize a 1-subspace, so that the action on V of either of these two subgroups is indecomposable. On the other hand, a direct sum decomposition $V = V_1 \oplus V_3$ of V has stabilizer K_2 isomorphic to $GL(V_3) \cong L_3(2)$. Since the possibility (1) has been exhausted, K_2 corresponds to (2). The description we have obtained is summarized in Table 3.1, where the second column gives the orbit lengths of K_is on Ω and the next two columns describe the structures of V and V^* as K_i-modules.

The subgroup K_2 acts on Ω doubly transitively. In terms of the $L_2(7)$ incarnation of K_2 the action is on the set of one-dimensional subspaces of the corresponding two-dimensional $GF(7)$-space. Since K_2 is contained both in a conjugate of M_1 and in a conjugate of M_3 we have the following *double complement principle.*

Lemma 3.10 *The following assertions hold:*

(i) *the semidirect product $M \cong 2^3 : L_3(2)$ of an elementary abelian group V_3 of order 8 and its automorphism group $K \cong GL(V_3) \cong L_3(2)$ contains precisely 16 subgroups isomorphic to $L_3(2)$, which constitute two M-conjugacy classes of size 8 each;*

(ii) *the $L_3(2)$-subgroups of M are conjugate in the automorphism group of M, which is a semidirect product of an elementary abelian group V_4 of order 16 and K;*

(iii) *the action of K on V_4 is indecomposable with a three-dimensional submodule V_3 and a one-dimensional factor-module; the action of K on the set of V_4-vectors outside V_3 is transitive just as on the set of eight Sylow 7-subgroups of K by conjugation.* □

The relationship between the two classes of complements when they are acting on a four-dimensional $GF(2)$-space is described by the following *switching principle.*

Lemma 3.11 *Let $M \cong 2^3 : L_3(2)$ act faithfully on a four-dimensional $GF(2)$-space $V_4(2)$ as a maximal parabolic subgroup. Let K_1 and K_2 be $L_3(2)$-subgroups from different conjugacy classes of M. Then the actions of K_1 and K_2 on $V_4(2)$ are non-isomorphic: one is decomposable and the other one is not.* □

We conclude this subsection by presenting the following $SL_2(7)$*-property* which will play an important role in the study of possible 2-groups extended by $L_3(2)$.

Lemma 3.12 *Let $K \cong L_3(2)$ act indecomposably on $V_4(2)$ stabilizing a one-dimensional subspace V_1. Let K^* be an $L_3(2)$-subgroup in $(V_4(2) : K)/V_1$ non-conjugate to KV_1/V_1. Then the pre-image of K^* in $V_4(2) : K$ does not split over V_1 and is isomorphic to $SL_2(7)$.* □

3.8.2 $S_6 \cong Sp_4(2)$

Let $\tau = (\alpha, \beta)$ be a transposition in $S(\Omega)$ and let $\Delta = \Omega \setminus \{\alpha, \beta\}$ be the complement in Ω to the support of τ. Then

$$C_A(\tau) = (\mathrm{Sym}(\Delta)\times\mathrm{Sym}(\{\alpha, \beta\}))^+ = \{(g, \tau^{\frac{1}{2}(\mathrm{sgn}(g)-1)}) \mid g \in \mathrm{Sym}(\Delta)\} \cong S_6.$$

We are going to associate with τ a symplectic polarity ρ of $V = V_4(2)$, which by definition is an isomorphism

$$\rho_\tau : V \mapsto V^*$$

such that v is contained in $\rho(v)$ for every $v \in V$. We follow the description of the projective geometry of $V_4(2)$ in terms of radicals of orders 8 and 16. Let $p \in \mathcal{P}$ be a point identified with its radical $E(p)$ of order 8. Since $E(p)$ acts regularly on Ω, there is a unique involution $r \in E(p)$ which maps α onto β. Then r is the only non-trivial element in $E(p)$ which commutes with τ (alternatively r can be defined as the unique element in $E(p)$ having the transposition τ in its decomposition into disjoint cycles). Then there is a unique hyperplane $\pi \in \Pi$ containing r in its radical $E(\pi)$. Then the required polarity ρ_τ sends p onto π. It is clear that $p \in \pi$ and the bilinear form f on V, defined in the usual manner by

$$f(p, q) = 0 \text{ if } q \in \rho_t(p) \text{ and } f(p, q) = 1 \text{ otherwise,}$$

is symplectic and it is manifestly non-degenerate.

Therefore $C_A(\tau) \cong Sp_4(2)$ and the claimed isomorphism is established. Certainly V is the natural symplectic module of $C_A(\tau) \cong Sp_4(2)$, which is self-dual and hence isomorphic to V^*. On the other hand, the heart U of the $GF(2)$-permutation module of $C_A(\tau) \cong S_6$ acting on Δ is also a four-dimensional irreducible module, which carries a non-degenerate symplectic form. In fact V and U are non-isomorphic $C_A(\tau)$-modules, and an isomorphism between them can be established only by an outer automorphism of S_6. In fact, a vector $r \in V \setminus \{0\}$ is a fixed-point free involution p on Ω with the transposition (α, β) removed, thus r corresponds to a partition of Δ into three disjoint pairs. On the other hand, a non-zero vector from U is a partition of Δ of the form $6 = 2 + 4$, and therefore corresponds to a pair from Δ.

3.8.3 $(S_3 \times S_5)^+ \cong \Gamma L_2(4)$

Let θ be a 3-cycle in $A(\Omega)$. Then θ acts on V fixed-point freely and hence endows it with the structure of a two-dimensional $GF(4)$-space. The normalizer in $A(\Omega)$ of the subgroup generated by θ is the stabilizer of this $GF(4)$-structure up to field automorphism. This is also relevant to the isomorphism between S_5 and $P\Gamma L_2(4)$.

By considering the above discussion in view of the outer automorphisms of S_6, we observe that A contains two conjugacy classes of subgroups A_5. A representative $A_5^{(1)}$ of one class commutes with a 3-cycle and fixes three elements of Ω, while a representative $A_5^{(2)}$ of the other class fixes two points and permutes transitively the rest. Here $A_5^{(1)}$ is contained in $(S_3 \times S_5)^+$, and $A_5^{(2)}$ is contained in $S_6 \cong Sp_4(2)$ and stabilizes a non-singular quadratic form of minus type. Thus we have the following A_5-*module description.*

Lemma 3.13 *The alternating group* A_5 *of degree* 5 *possesses precisely two faithful four-dimensional* $GF(2)$*-modules: the* linear module *where* A_5 *acts transitively on the set of non-zero vectors and preserves a* $GF(4)$*-structure, and the* orthogonal module *on which* A_5 *preserves a quadratic form of minus type and has two orbits on the non-zero vectors with lengths* 5 *and* 10*, comprised by isotropic and non-isotropic vectors, respectively.* □

3.9 Symplectic Identification

In this section we show that the subgroups $S_6 \cong Sp_4(2)$ in $A \cong A_8 \cong L_4(2)$ are characterized as subgroups generated by certain subamalgams. Let $\mathcal{S} = \{P_1, P_2\}$ be Goldschmidt's amalgam G_3^1. This is a coreless amalgam with $P_1 \cong P_2 \cong S_4 \times 2$ and $P_{12} := P_1 \cap P_2 \cong D_8 \times 2$. It was mentioned by Goldschmidt himself that $P = S_6 \cong Sp_4(2)$ is a completion group of $\mathcal{S}$. This can be seen as follows. As above, let Ω be an 8-set of which A is the alternating group and let Δ be a 6-subset in Ω, whose stabilizer P in A is $\mathrm{Sym}(\Delta) \cong S_6 \cong Sp_4(2)$. Let P_1 be the stabilizer of a partition of Δ into three two-element subsets and let P_2 be the stabilizer of a 2-subset from that partition.

In the symplectic incarnation let $(V_4(2), f)$ be a four-dimensional $GF(2)$-space on which A acts as the full linear group, and let f be the symplectic form constructed in Subsection 2.7.3, so that $P \cong Sp_4(2) = \mathrm{Aut}(V_4(2), f)$. Let P_1 be the stabilizer of a non-zero vector in $V_4(2)$ and let P_2 be the stabilizer of a singular two-dimensional subspace containing the vector stabilized by P_1. Then $\mathcal{S} = \{P_1, P_2\}$ is a subamalgam in A isomorphic to G_3^1 which generates P.

Let us classify the subamalgams in $A \cong A_8 \cong L_4(2)$ isomorphic to G_3^1 to demonstrate that up to conjugation there is a unique such subamalgam, thus generating $P \cong S_6 \cong Sp_4(2)$. Let $\mathcal{S} = \{P_1, P_2\}$ be such a subamalgam, and let z_1 and z_2 be the non-trivial elements in $Z(P_1)$ and $Z(P_2)$, respectively. For $i = 1$ and 2 let $X(P_i)$ be the largest elementary abelian normal subgroup of P_i, let N_i be a complement to $X(P_i)$ in P_i, and let $Y(P_i)$ be the commutator of N_i in $X(P_i)$. Then

$$X(P_i) \cong 2^3, \ N_i \cong S_3, \ Y(P_i) \cong 2^2,$$

and $Y(P_i)$ is the Klein 4-subgroup of an S_4-direct factor of P_i. Furthermore, z_1 and z_2 are non-squares in $Z(P_{12}) \cong Z(D_8 \times 2) \cong 2^2$. The classification will be accomplished in the following steps: (1) identify z_1 and z_2 among involutions in A; (2) choose P_1 and P_2 as $(S_4 \times 2)$-subgroups in $C_A(z_1)$ and $C_A(z_2)$, respectively; and (3) check that $P_1 \cap P_2 \cong D_8 \times 2$ and that only the identity subgroup of P_{12} is normal in both P_1 and P_2.

There are two classes of involutions in A: one consists of central involutions (in the sense that they can be found in the centre of a Sylow 2-subgroup of A) and the other consists of non-central involutions.

Non-central involutions. If n is a non-central involution in A then it acts on Ω as the product of two transpositions,

$$C_A(n) = (D_8 \times S_4)^+ \cong (2^2 \times A_4).2,$$

where as before the superscript $+$ indicates the intersection with the alternating group. This shows that $C_A(n)$ contains a single $(S_4 \times 2)$-subgroup, which is normal. The centre of a Sylow 2-subgroup in such a $(S_4 \times 2)$-subgroup contains one central involution and two non-central involutions.

Central involutions. If c is a central involution in A then it acts on Ω fixed-point freely and it acts on $V_4(2)$ as a transvection. From the latter viewpoint let v be the centre of c and V_3 be its axis, so that $c = \tau(v, V_3)$. Let E_1 and E_3 be the radicals of order 8 in the stabilizers in A of v and V_3, respectively. Let w

be a vector outside V_3, let V_2 be a 2-subspace in V_3 not containing v, and let $N \cong S_3$ be the largest subgroup in A which stabilizes the 2-subspace V_2 as a whole and the 2-subspace $\langle v, w \rangle$ vector-wise. Then

$$C_A(c) = E_1 E_3 N \cong 2_+^{1+4} : S_3,$$

where 2_+^{1+4} denotes the extraspecial group of order 32 of plus type.

Let $B = (b_1, b_2, b_3, b_4)$ be a basis of $V_4(2)$ such that $b_1 = v$, $\langle b_1, b_2, b_3 \rangle = V_3$, $\langle b_2, b_3 \rangle = V_2$ and $w = b_4$, and let τ_{ij} be the transvection with centre b_i and axis $\langle b_k \mid 1 \le k \le 4, b_k \ne j \rangle$. In this basis the transvection τ_{ij} is represented by a matrix with 1s on the diagonal and the only off-diagonal non-zero entry in the ith row and jth column. Then the matrix realization of $C_A(c)$ in the basis B can be expressed as follows:

$$C_A(c) = \begin{pmatrix} 1 & 0 & 0 & 0 \\ \tau_{21} & & & 0 \\ & & S_3 & \\ \tau_{31} & & & 0 \\ \tau_{41} & \tau_{42} & \tau_{43} & 1 \end{pmatrix}.$$

In the above expression the S_3 placed in the centre of the matrix symbolizes the Levi complement $N \cong S_3$, and the transvection τ_{ij} is positioned in the only off-diagonal non-zero entry in its matrix realization.

There are three N-invariant elementary abelian subgroups of order 8 in $C_A(c)$, namely E_1, E_3 and

$$D = \langle c, \tau(u, V_3) \cdot \tau(v, \langle v, u, w \rangle) \mid u \in V_2 \setminus \{0\} \rangle = \langle \tau_{41}, \tau_{31}\tau_{42}, \tau_{21}\tau_{43} \rangle$$

where 'D' stands for *diagonal*. It follows immediately that $C_A(c)$ contains three $(S_4 \times 2)$-subgroups containing $\langle c, N \rangle$, namely

$$C_1 = E_1 : N, \; C_2 = D : N, \; C_3 = E_3 : N.$$

If $P_1 = C_2$ then the centre of a Sylow 2-subgroup in P_1 contains a unique involution z_2 different from $c = z_1$, which is not a square, and this involution is non-central. If we set P_2 to be the unique $(S_4 \times 2)$-subgroup in $C_A(z_2)$ then we do indeed obtain a copy of the G_3^1-amalgam in A. On the other hand, if P_1 is chosen to be C_1 or C_3 then the centre of a Sylow 2-subgroup in P_1 contains only central involutions. Therefore $X(P_2)$ also contains only central involutions and hence it must be a radical of order 8 intersecting the radical $X(P_1)$ in an order-4 subgroup. This enforces the equality $X(P_1) = X(P_2)$, which contradicts the condition that $\mathcal{S}$ must be core-free. Thus we have established the following *symplectic identification*.

Lemma 3.14 *Let $\mathcal{S} = \{P_1, P_2\}$ be a core-free subamalgam in $A \cong A_8 \cong L_4(2)$ such that $P_1 \cong P_2 \cong S_4 \times 2$ and $[P_1 : P_{12}] = [P_2 : P_{12}] = 3$, where $P_{12} = P_1 \cap P_2$. Then $\mathcal{S}$ is unique up to conjugation in A and it generates a subgroup $P \cong S_6 \cong Sp_4(2)$.* □

3.10 Restricting Automorphisms of $2^3 : L_3(2)$

Let $M \cong 2^3 : L_3(2)$ be the semidirect product of $V_3 \cong 2^3$ and $L \cong L_3(2) \cong \mathrm{Aut}(V_3)$. By the *double complement* Lemma 3.10 the group $\mathrm{Aut}(M)$ is a semidirect product of $V_4 \cong 2^4$ and L with L acting on V_4 indecomposibly: it stabilizes V_3 and acts fixed-point freely on $V_4 \setminus V_3$. Therefore, every element $u \in V_4 \setminus V_3$, when acting via conjugation, induces an outer automorphism of M possessing the following features:

(1) $[u, V_3] = 1$;
(2) L^u is not conjugate to L in M;
(3) u centralizes LV_3/V_3;
(4) $C_M(u) \cong F_{21}$; in particular,
(5) u centralizes seven subgroups of order 3, but acts fixed-point freely on the set of involutions in M which are not in V_3.

3.10.1 Point Stabilizer

Let $v \in V_3$ and let $D = C_M(v) \cong 2^3 : S_4$. Since $V_3 \leq D$, by (3) we have $D^u = D$ and by Sylow's theorem we can choose u to centralize a Sylow 3-subgroup T in D. We shall demonstrate how the properties (1) to (5) and the structure of D enable us to reconstruct uniquely the action of u on D.

Let $U_1 = [V_3, T]$ and $U_2 = O_2(C_L(v))$. Then every non-identity element from U_2 is a transvection $\tau(w)$ with centre v and axis $\langle v, w\rangle$ for some $w \in U_1$. Therefore there is a natural bijection between U_1 and U_2 which sends w to $\tau(w)$. The subgroup $Q := O_2(D)$ is generated by v, U_1 and U_2, and it is easy to see from the above that Q is extraspecial of order 2^4, of $+$ type, with centre $\langle v\rangle$, and that T acts on $\bar{Q} := Q/\langle v\rangle$ fixed-point freely. Let ι be an involution in L which normalizes T. Then ι is contained in D, since $S_3 = \langle T, \iota\rangle$ is the full normalizer of T in L. Notice that v, U_1, U_2, T and ι generate D.

The fixed-point free action of T on $\bar{Q}$ induces on the latter a $GF(4)$-vector space structure on which ι acts as a field automorphism. There are five one-dimensional subspaces with respect to this structure; the pre-images in Q of two of them are quaternion groups and they are permuted by ι, while the remaining three are normalized by ι and have elementary abelian pre-images. This is easy to deduce from the isomorphism between S_5 and the extension of $PGL_2(4)$ by a field automorphism.

The two one-dimensional $GF(4)$-subspaces normalized by ι are the images of U_1 and U_2, while the third one is the image of

$$U_3 = \langle (w + v)\tau(w) \mid w \in U_1^{\#} \rangle,$$

(where commonly $X^{\#}$ denotes the set of non-zero vectors of a vector space X). By (3) and (5) $\iota^u = \iota x$, where $x \in V_3 \setminus \{0\}$ and x is centralized by T, which forces x to be v. Since the actions of ι and ιu on Q are the same, by (5), u must move U_2 onto a similar subgroup normalized by T and ι, and this subgroup can only be U_3. Since the commutator relations with the elements of U_1 must be preserved, we have

$$u : \tau(w) \mapsto \tau(w)w$$

for $w \in U_2^{\#}$, and by this information the action of u on D has been completely reconstructed.

It would appear appropriate to make the following remark. The subgroup T as described above commutes with two involutions in $V_4 \setminus V_3$: if u is one of them then the other one is uv. Since v is in the centre of D, the actions on D of u and uv coincide, so that the uniqueness conclusion is consistent with the double choice of u.

3.10.2 Line Stabilizer

Next, let V_2 be a 2-subspace in V_3, and let $E = N_M(V_3) \cong 2^3 : S_4$. Similarly to the previous case, we choose a Sylow 3-subgroup $T \in E \cap L \cong S_4$. For an element $u \in V_4 \setminus V_3$ commuting with T we aim to reconstruct the possibilities for the action of u on E. Let v be the unique non-identity element in V_3 commuting with T. Then v is not in the centre of E and, unlike in the previous case, we should expect at least two possibilities for the action: one corresponding to u and the other to uv.

If we put $W_1 = V_2$, $W_2 = O_2(N_L(V_2))$ and $W = W_1 W_2$, then

$$O_2(E) = W \times \langle v \rangle = [O_2(E), T] \times C_{O_2(E)}(T).$$

The subgroup W is elementary abelian of order 2^4, on which T acts fixed-point freely, thus inducing on W the structure of a two-dimensional $GF(4)$-space. The subgroups W_1 and W_2 of order 4 are normalized by T, and therefore they correspond to 1-subspaces with respect to the $GF(4)$-structure on W. Let W_3, W_4 and W_5 be the remaining 1-subspaces, and let $\mathcal{P}(W)$ denote the set of all five 1-subspaces. If $\tau(w)$ denotes the transvection with centre w and axis V_2, then we assume that

$$W_3^\# = \{w\tau(w) \mid w \in V_2^\#\}.$$

If ι is an involution in form $N_L(T) \cong S_3$, then ι acts on W as a field automorphism. By the choice of our notation ι normalizes W_1, W_2 and W_3, and hence ι acts on $\mathcal{P}(W)$ as the transposition (W_4, W_5) from

$$\mathrm{Sym}(\mathcal{P}(W)) \cong S_5 \cong PGL_2(4).\langle\iota\rangle.$$

The element v acts on $\mathcal{P}(W)$ as the permutation $(W_1)(W_2\ W_3)(W_4\ W_5)$ (which is even, because v is contained in $A_5 \cong PGL_2(4)$). Therefore, E is a semidirect product of $W \cong 2^4$ and $T\langle\iota\rangle \times \langle v\rangle \cong S_3 \times 2$, where the action is as described above.

Let us look at the possibilities for the action of u. Since u does not centralize involutions in $E \setminus V_3$, and since u normalizes $N_E(T) = T\langle\iota\rangle \times \langle v\rangle$, we deduce the following crucial equality which we call the *product equality*.

$$\iota^u = \iota v.$$

The action of u on $\mathcal{P}(W)$ is an even permutation (since u commutes with T), which commutes with the action of v. Thus u acts either as $(W_1)(W_2\ W_4)(W_3\ W_5)$, or as $(W_1)(W_2\ W_5)(W_3\ W_4)$. Since

$$GL(W) \cong GL_2(4) \cong 3 \times A_5,$$

each of the above two actions uniquely extends to an action on the whole of W, which is consistent with the choice between u and uv.

3.11 Projective Plane of Order 4

Let $V_6(2)$ be a six-dimensional $GF(2)$-space and let T be a fixed-point free subgroup of order 3 in $GL(V_6(2))$. Then T, together with the zero endomorphism, will constitute in the endomorphism ring of $V_6(2)$ the subring isomorphic to the field $GF(4)$ of order 4. Therefore,

$$C_{GL(V_6(2))}(T) \cong GL(V_3(4)) \cong GL_3(4)$$

preserves on $V_6(2)$ the structure of a three-dimensional $GF(4)$-space $V_3(4)$. An involution ι from $GL(V_6(2))$ which inverts T fixes a basis of $V_3(4)$, and hence acts on this space as a field automorphism. The group T acts on $V_3(4)$ by scalar multiplication and

$$GL(V_3(4)).\langle\iota\rangle/T \cong P\Gamma L_3(4)$$

is the automorphism group of the projective plane $\Xi = (\mathcal{P}_4, \mathcal{L}_4)$ of order 4, where $\mathcal{P}_4$ and $\mathcal{L}_4$ are 1- and 2-subspaces in $V_3(4)$, and the incidence relation is via inclusion. Notice that $GL(V_3(4))$ does not split over T.

A projective planes of order 4 is unique subject to the following standard axioms:

(Ξ1) $|\mathcal{P}_4| = |\mathcal{L}_4| = 21$;
(Ξ2) every point is on five lines and there are five points on every line;
(Ξ3) any two distinct points are on a unique common line and any two lines intersect in a single point.

As usual, when convenient we identify a line with its point-set. Recall that $PSL_3(4) = L_3(4)$ is another simple group of order $\frac{1}{2}8!$, which is not isomorphic to the simple group of that order discussed earlier, which occurs in $L_4(2)$ and A_8 incarnations. The group $P\Gamma L_3(4)$ is a subgroup in the automorphism group of $L_3(4)$, which splits over $L_3(4)$ with complement isomorphic to S_3. To obtain the full automorphism group of $L_3(4)$ one should adjoin the inverse–transpose automorphism, which establishes an isomorphism of Ξ and its dual.

In the rest of this section we discuss some familiar features of Ξ, namely the Fano subplanes and the hyperovals. A set of points in $\mathcal{P}_4$ is said to be *independent* if every line intersects it in at most two points. An independent 3-set is just a basis of the underlying $GF(4)$-space, where vectors are considered up to their non-zero scalar multiples. If the basis is chosen to be standard then the stabilizer D of such a 3-set is comprised of the diagonal matrices, and in order to form an independent 4-set one should adjoin a vector whose coordinates are all non-zero. Since this vector is determined up to scalar multiples, we arrive at nine possibilities. If D is taken from $GL_3(4)$ then the action on the nine candidates is transitive, but if it is from $SL_3(4)$ then there are three orbits, each of size 3. Therefore there are

$$2520 = \frac{21 \cdot 20 \cdot 16 \cdot 9}{4!}$$

independent 4-sets in Ξ that are transitively permuted by $PGL_3(4)$ or $P\Gamma L_3(4)$, each of which splits into three orbits of equal size under the action of $L_3(4)$. We are going to establish the following *Fano–hyperoval uniqueness* result.

Lemma 3.15 *Every independent* 4*-set in* Ξ *is contained in a unique maximal independent set of size* 6 *called a hyperoval and in a unique Fano subplane.*

Proof Let X be an independent quadruple of points in Ξ, and let $L(X)$ be the set of six lines intersecting X in two points. Then the set of pairwise intersections of the lines in L_X contains X and three further points. Let Y be the set of these three new points (corresponding to the partitions of X into disjoint pairs). Then it turns out that *Y is contained in a line $l(Y) \in \mathcal{L}_4$*. Then the hyperoval and the subplane are

$$X \cup l(Y) \setminus Y \quad \text{and} \quad (X \cup Y, L(X) \cup \{l(y)\}).$$

□

A Fano subplane can also be obtained as the set of points and lines in Ξ fixed by a field automorphism. This observation in view of the previous discussions provides us with the following.

Lemma 3.16 *The projective plane Ξ of order 4 contains $360 = 2520/7$ Fano subplanes. The set of Fano subplanes is transitively permuted by $P\Gamma L_3(4)$ and splits into three orbits, each of size 120, under the action of $L_3(4)$. The stabilizer of a subplane Φ in $P\Gamma L_3(4)$ induces on Φ its full automorphism group $L_3(2)$ with kernel of order 2, which splits as a direct factor. The intersection of the point-sets of two subplanes is odd or even depending on whether they are taken from the same or from different $L_3(4)$-orbits.* □

The numerology of the *hyperovals* is similar.

Lemma 3.17 *The projective plane Ξ of order 4 contains $168 = 2520/15$ hyperovals. The set of hyperovals is transitively permuted by $P\Gamma L_3(4)$, while it splits into three orbits, each of size 56, under the action of $L_3(4)$. The stabilizer of a hyperoval Θ in $P\Gamma L_3(4)$ acts faithfully on Θ and induces the full symmetric group Sym$(\Theta) \cong S_6$. The pre-image of the stabilizer in $\Gamma L_3(4)$ does not split over the centre of $GL_3(4)$ of order 3. The size of the intersection of two hyperovals is even or odd, depending on whether they are taken from the same or different $L_3(4)$-orbits. A subplane and a hyperoval intersect evenly if and only if their $L_3(4)$-orbits have the same stabilizer in $P\Gamma L_3(4)$.* □

The group $P\Gamma L_3(4) = \mathrm{Aut}(\Xi)$ contains a single class of involutions outside $L_3(4)$, so that every such involution ι is a field automorphism. Thus the set $\Phi(\iota)$ of points in $\mathcal{P}_4$ fixed by ι makes (the point-set of) a Fano subplane. Let $l \in \mathcal{L}_4$ be a line, which ι stabilizes as a set. Then ι fixes three points on l (called the *horns*) and permutes the remaining two points (called the *hooves*). Then the set of horns of the lines stabilized by ι makes up the line-set of $\Phi(\iota)$.

The above construction of a hyperoval and a subplane from an independent 4-set demonstrates the following *horn–hoof principle*.

Lemma 3.18 *Let Φ be a Fano subplane and let l be a line from $\mathcal{P}_4$ having three points in Φ (the horns) and two points outside Φ (the hooves). Then, by removing the horns l and adjoining its hooves, we obtain a 6-set of points which is a hyperoval.* □

With ι being a field automorphism and $\Phi(\iota)$ being the subplane formed by the points fixed by ι, let J be the centralizer of ι in $P\Gamma L_3(4) = \mathrm{Aut}(\Xi)$. Then J is the stabilizer of $\Phi(\iota)$ in the latter group and

$$J = L \times \langle \iota \rangle \cong L_3(2) \times \langle \iota \rangle \cong \mathrm{Aut}(\Phi(\iota)) \times \langle \iota \rangle.$$

Let τ be an involution in L. Then we know that τ acts on $\Phi(\iota)$ as a transvection, whose centre we denote by p and whose axis is l (clearly $p \in l$). Then $\tau\iota$ is again a field automorphism and the relationship between $\Phi(\iota)$ and $\Phi(\tau\iota)$ is dictated by the *flag reflection property*.

Lemma 3.19 *In the above terms $\Phi(\iota) \cap \Phi(\tau\iota)$ is the set of the three horns of l, while $\Phi(\iota) \cup \Phi(\tau\iota)$ is the union of three lines from $\mathcal{L}_4$ passing through p with the hooves of l removed.* □

In order to justify this we recall that the stabilizer of a line (viewed as a set of five points) in $L_3(4)$ induces on this line the alternating group A_5 all of whose involutions have one fixed point each, while the similar stabilizer in $P\Gamma L_3(4)$ induces the symmetric group S_5, whose extra class of involutions consists of the actions of field automorphisms. We will say that $\Phi(\iota)$ and $\Phi(\tau\iota)$ are obtained from each other by *reflection with respect to the flag* (p, l).

The following procedure which transfers a Fano subplane into a hyperoval will play a crucial role. We call it the *transfer principle*.

Lemma 3.20 *Let Φ be a Fano subplane in Ξ, let $p \in \Phi$ be a point and let l_1, l_2 and l_3 be the lines of Φ passing through p. Then the set $\Theta(\Phi, p)$ of six hooves of the l_is is a hyperoval.* □

3.12 Outer Automorphism of S_6

On looking at the projective plane $\Xi = (\mathcal{P}_4, \mathcal{L}_4)$ through a hyperoval one comes up with a nice model of the plane. Considering a hyperoval P_1 as an abstract 6-element set, define $L_1 = \binom{P_1}{2}$ to be the set of 2-element subsets of P_1, $P_2 = \binom{P_1}{2|2|2}$ to be the set of partitions of P_1 into three pairs, and L_2 to be the set of maximal subsets of P_2 subject to the condition that any two partitions from a set involve different pairs. It is well known and easy to check that there are six subsets in L_2, each of size 5. Put $\mathcal{P}_4 = P_1 \cup P_2$ and $\mathcal{L} = L_1 \cup L_2$, and define the incidence relation via inclusion. Then

$$|\mathcal{P}_4| = |P_1| + |P_2| = 6 + 15 = 21, \quad |\mathcal{L}_4| = |L_1| + |L_2| = 15 + 6 = 21,$$

and all the defining properties of $\Xi = (\mathcal{P}_4, \mathcal{L}_4)$ are readily verifiable. Manifestly every permutation from

$$X := \mathrm{Sym}(P_1) \cong S_6$$

extends uniquely to an automorphism of Ξ and, by the order consideration or otherwise, X is the stabilizer of the hyperoval P_1 in $\mathrm{Aut}(\Xi)$. Furthermore, L_2 is a hyperoval in the dual plane $\Xi^* = (\mathcal{L}_4, \mathcal{P}_4)$, so that X is also the stabilizer of L_2 in $\mathrm{Aut}(\Xi^*) = \mathrm{Aut}(\Xi)$. Moreover, $\Xi^* \cong \Xi$ because of the uniqueness of the projective plane of order 4 and, since $\mathrm{Aut}(\Xi)$ acts transitively on the set of hyperovals in Ξ, there is an isomorphism

$$\iota\colon (\mathcal{P}_4, \mathcal{L}_4) \to (\mathcal{L}_4, \mathcal{P}_4),$$

such that $\iota(P_1) = L_2$. By the above, X must be normalized by ι, and we claim that the automorphism of X it induces is *not* an inner one. In fact, ι clearly maps P_2 onto L_1, where the former is indexed by the transpositions in $\mathrm{Sym}(P_1)$, while the latter is indexed by the products of three commuting transpositions. This leads us to the following *outer S_6-automorphism* property.

Lemma 3.21 *Let P_1 be a set of six elements and let $X \cong S_6$ be the symmetric group of P_1. Let K_2 be the class of involutions in X which act on P_1 as transpositions and let $K_{2|2|2}$ be the class of involutions in X which act on P_1 fixed-point freely. Then X possesses an (outer) automorphism ι which permutes K_2 and $K_{2|2|2}$.* □

It is worth mentioning that $\mathrm{Aut}(S_n) = S_n$ if $n \geq 3$ and $n \neq 6$, while $\mathrm{Out}(S_6)$ is of order 2. If $Y^{(0)}$ is the commutator subgroup of X isomorphic to A_6 and

$$Y = \mathrm{Aut}(X) = \mathrm{Aut}(Y^{(0)})$$

then there are three proper subgroups in Y which properly contain $Y^{(0)}$. One of these three subgroups is $X \cong S_6$, another one is isomorphic to $PGL_2(9)$ and the last one is known as the Mathieu group of degree 10 (denoted by M_{10}). The group M_{10} can be specified by the property of having no involutions outside $Y^{(0)}$. The latter group is the point stabilizer in the natural action of the smallest Mathieu group, M_{11}, on 11 points.

4

From $L_5(2)$ to the Mathieu Amalgam

In this chapter we start by establishing the notation for the (projective) general linear group $H = L_5(2)$ in dimension 5 over the field $GF(2)$ of two elements. We construct the subamalgam $\mathcal{H}$ in H formed by the stabilizers of pairwise incident one-, two- and three-dimensional subspaces in the natural module. Then $\mathcal{H}$ is precisely the $L_5(2)$-amalgam we will *twist* to obtain the Mathieu amalgam $\mathcal{A}$.

4.1 Maximal Parabolics in $GL_n(2)$

Let V be an n-dimensional $GF(2)$-space, where $n \geq 2$, and let

$$L = GL(V) \cong GL_n(2) \cong L_n(2)$$

be the general linear group of V. Notice that V can be viewed as an elementary abelian group of order 2^n, while L can be viewed as the automorphism group of V. Let U be an i-dimensional subspace in V, where $1 \leq i \leq n-1$, and let W be a complement to U in V, so that

$$V = U \oplus W.$$

The stabilizer L_i of U in L (known as a maximal parabolic subgroup of L) possesses the following structural description.

Recall that, if

$$0 \neq v \in X < V$$

and dim $(X) = n-1$, then the *transvection* $\tau(v, X)$ with centre v and axis X is an element of L which fixes X vector-wise and adds v to every vector outside X. The transvections $\tau(v, X)$ subject to

$$v \in U \leq X$$

are contained in L_i, pairwise commute and generate the *radical* R_i of L_i, which is the kernel of L_i acting on $U \oplus V/U$. The generation of R_i by transvections that commute (see Proposition 4.1 below) immediately demonstrates that it is elementary abelian of rank $i \cdot (n-i)$. The radical R_i is complemented in L_i by the stabilizer K_i (known as a *Levi complement*) of the direct sum decomposition $V = U \oplus W$, so that

$$L_i = R_i : K_i.$$

Furthermore, K_i is the direct product of the kernel K_i^- of its action on the set of subspaces contained in U (thus on the whole of U) and the kernel K_i^+ of its action on the set of subspaces containing U (equivalently on $W \cong V/U$), so that

$$K_i = K_i^+ \times K_i^- \cong GL(U) \times GL(W) \cong GL_i(2) \times GL_{n-i}(2).$$

By conjugating a generating transvection $\tau(v, X)$ of R_i by an element of $k \in K_i$ with the direct product components k^+ and k^- we obtain another generating transvection, $\tau(k^+(v), k^-(X))$, where $k^+(v)$ has an obvious meaning, while $k^-(X)$ is obtained through the isomorphism between W and V/U, which commutes with K. This action, when extended onto the whole of R_i, defines on the latter the tensor product structure $U \otimes W^*$, where W^* is the module dual to W. Thus

$$L_i \cong (U \otimes W^*) : (GL(U) \times GL(W)).$$

Denoting by 2^m (as commonly accepted) the elementary abelian group of that order, we obtain

$$L_i \cong 2^{i(n-i)} : (GL_i(2) \times GL_{n-i}(2)),$$

where the tensor product structure on $R_i \cong 2^{i(n-i)}$ is implicit. It is worth mentioning that $L_i = N_L(R_i)$ for all is and this is a key observation for building up the structure of $V_n(2)$ from $GL_n(2)$ when the latter is given as an abstract group. This was fully accomplished in the case $n = 4$ in Lemma 3.9, where $L_4(2)$ was given as the alternating group on eight points.

4.2 Five-dimensional $GF(2)$-Space

Let $V = V_5(2)$ be a five-dimensional $GF(2)$-space, and let V^* be the space dual to V. Let $B = (b_1, \ldots, b_5)$ be a basis of V and let $B^* = (b_1^*, \ldots, b_5^*)$ be the dual basis, so that, viewing b_i^* as a linear function on V, we have

$$b_i^*(b_j) = \delta_{ij}$$

(the Kronecker delta). Since the ground field is of order 2, a vector from V^* will be identified with its kernel (when considered as a linear function on V). The vectors from V are represented by columns formed by the coordinates in B, and the vectors from V^* are represented by rows formed by the coordinates in B^*. Therefore, the natural evaluation map

$$V^* \otimes V \to GF(2)$$

is performed by the usual scalar product, while the elements of the linear group

$$H := GL(V) = GL(V^*) \cong L_5(2)$$

are realized by matrices acting on V from the left and on V^* from the right.

For $1 \leq i \neq j \leq 5$ the transvection $\tau_{ij} = \tau(b_i, b_j^*)$ having b_i as the centre and b_j^* as the axis fixes the vectors in B, except for b_j, which it sends onto $b_j + b_i$. Thus τ_{ij} is represented by the matrix which (a) has ones on the diagonal; and (b) has zeros everywhere else; except for (c) the entry in the ith row and jth column, which is one. The product of two transvections is described in the following *transvection product* rule, which is well known, easy to check and holds for all n.

Proposition 4.1 *If $\tau_{ij} = \tau(b_i, b_j^*)$ and $\tau_{kl} = \tau(b_k, b_l^*)$ are distinct transvections and $t = \tau_{ij}\tau_{kl}$ then one of the following holds:*

(i) *$i = k$ or $j = l$, $[\tau_{ij}, \tau_{kl}] = 1$ and $t = \tau(b_i, b_j^* + b_l^*)$ or $t = \tau(b_i + b_k, b_j^*)$, respectively;*
(ii) *$i = l$ and $j = k$, and t is an element of order 3;*
(iii) *$i = l$ with $j \neq k$ or $j = k$ with $i \neq l$, and t is of order 4 with $t^2 = \tau_{kj}$ or $t^2 = \tau_{il}$, respectively;*
(iv) *$\{i, j\}$ and $\{k, l\}$ are disjoint sets, $[\tau_{ij}, \tau_{kl}] = 1$ and t is not a transvection.*

□

We need a pair of opposite flags in V and a pair of opposite flags in V^*, which are dual to each other. Let U_i be the subspace in V spanned by the first i basis vectors and V_i be the subspace spanned by the last i basis vectors, so that

$$V = U_{5-i} \oplus V_i$$

for $1 \leq i \leq 5$ and hence the following two flags are opposite to each other in the sense that $V_i \cap U_{5-i} = \{0\}$:

$$\Phi: \quad 0 < V_1 < V_2 < V_3 < V_4 < V_5 = V,$$
$$\Psi: \quad 0 < U_1 < U_2 < U_3 < U_4 < U_5 = V.$$

For $i = 1$, 2 and 3 let H_i be the stabilizer of V_i in H and let $\mathcal{H} = \{H_1, H_2, H_3\}$ be the subamalgam in H (called the *$L_5(2)$-amalgam*) formed by these subgroups. Since

$$V_1 = \langle b_5 \rangle, \quad V_2 = \langle b_4, b_5 \rangle, \quad V_3 = \langle b_3, b_4, b_5 \rangle,$$

the matrix forms of the members of the $L_5(2)$-amalgam are as illustrated below, where τ_{ij} is written in the ijth position to indicate the non-zero non-diagonal entry of the transvection in its matrix presentation:

$$H_1 = \left(\begin{array}{cccc|c} & & & & 0 \\ & \multicolumn{2}{c}{GL(U_4)} & & 0 \\ & & & & 0 \\ & & & & 0 \\ \tau_{51} & \tau_{52} & \tau_{53} & \tau_{54} & 1 \end{array}\right),$$

$$H_2 = \left(\begin{array}{ccccc} & & & 0 & 0 \\ & GL(U_3) & & 0 & 0 \\ & & & 0 & 0 \\ \tau_{41} & \tau_{42} & \tau_{43} & \multicolumn{2}{c}{\multirow{2}{*}{GL(V_2)}} \\ \tau_{51} & \tau_{52} & \tau_{53} & & \end{array}\right),$$

$$H_3 = \left(\begin{array}{ccccc} \multicolumn{2}{c}{GL(U_2)} & 0 & 0 & 0 \\ & & 0 & 0 & 0 \\ \tau_{31} & \tau_{32} & & & \\ \tau_{41} & \tau_{42} & & GL(V_3) & \\ \tau_{51} & \tau_{52} & & & \end{array}\right).$$

We will adopt the notations R_i, K_i, K_i^+ and K_i^- introduced in the previous section, so that

$$H_1 \cong R_1 : K_1 \cong 2^4 : L_4(2),$$
$$H_2 \cong R_2 : (K_2^+ \times K_2^-) \cong 2^6 : (L_3(2) \times S_3),$$
$$H_3 \cong R_3 : (K_3^+ \times K_3^-) \cong 2^6 : (S_3 \times L_3(2)).$$

One can notice that we prefer to write $L_n(2)$ rather than $GL_n(2)$, and S_3 rather than $L_2(2)$. The set of transvections generating the radicals R_i is clear from the above matrices. Of course H_2 and H_3 are isomorphic, but in our game they play totally different roles.

4.3 The Universal Completion of the $L_5(2)$-Amalgam

We have introduced $\mathcal{H} = \{H_1, H_2, H_3\}$ as a subamalgam in $H \cong L_5(2)$. Let us view it as an abstract amalgam. Then the third member H_3 can be recovered from the subamalgam $\{H_1, H_2\}$ in the following way. First, the radical R_3 can be specified inside $H_{12} = H_1 \cap H_2$ as the only elementary abelian subgroup of order 2^6 in a Sylow 2-subgroup of H_{12}, which is distinct from R_2. Second, H_3 is generated by the subamalgam $\mathcal{H}_3 := \{H_{13} = N_{H_1}(R_3), H_{23} = N_{H_2}(R_3)\}$. If C_3 is the core of the subamalgam $\mathcal{H}_3$ then $C_3 = R_3 : K_3^+ \cong 2^6 : S_3$ and H_3 is the only completion of $\mathcal{H}_3$ in which the core has a trivial centralizer. In fact, C_3 is R_3 extended by $K_3^+ \cong S_3$, if t is an element of order 3 from K_3^+ then t acts on R_3 fixed-point freely, thus defining on R_3 a $GF(4)$-vector space structure; an involution ι from K_3^+ inverts t and hence acts on R_3 as a $GF(4)$-field automorphism. One can take $t = \tau_{12}\tau_{21}$ and $\iota = \tau_{21}$. Then

$$\mathrm{Aut}(C_3)/C_3 \cong L_3(2),$$

while the images of H_{13} and H_{23} are two maximal parabolic subgroups in that $L_3(2)$ intersecting in a Sylow 2-subgroup, which is the image of H_{123}. Thus H_3 is the only completion of $\mathcal{H}_3$ in $\mathrm{Aut}(C_3)$ and this completion is surjective.

It is natural to ask whether we can recover H from $\mathcal{H}$ viewed as an abstract amalgam. The answer is in the affirmative and given in the following proposition.

Proposition 4.2 *The group $L_5(2)$ is the universal completion of the amalgam $\mathcal{H}$.*

Proof We reduce the identification with $L_5(2)$ of the universal completion X of $\mathcal{H}$ to the characterization of the projective $GF(2)$-geometries in Proposition 3.6 by accomplishing the following steps.

(X1) Set $\mathcal{P}$ to be the set of cosets of H_1 in X with p denoting the identity coset.

(X2) Show that the H_2-orbit l of p has length $3 = [H_2 : H_{12}]$ and that H_2 induces S_3 on l.

(X3) Define a graph Γ on $\mathcal{P}$ where two distinct cosets are adjacent if there is an X-element which maps them simultaneously into l.

(X4) Observe that Γ is undirected and connected, and that the subgraph on l is complete (a triangle).

(X5) There are $30 = [H_1 : H_{12}]$ edges containing p, which are transitively permuted by H_1, with R_1 having 15 orbits of length 2 indexed by the hyperplanes of R_1.

(X6) The H_3-orbit π of p has length $7 = [H_3 : H_{13}]$, $l \in \pi$ and H_3 induces on π the action of $L_3(2) = H_3/T_3$ as on the points of the associated Fano plane.

(X7) The subgraph on π is complete, and by (X5) so is the whole graph Γ, and hence $|\mathcal{P}| = 31$.

(X8) If $\mathcal{L}$ and Π are the sets of images under X of l and π, respectively, then the conditions (i) to (iii) in Proposition 3.6 hold. □

4.4 Automorphisms of H_{12}

We shall modify the $L_5(2)$-amalgam $\mathcal{H}$ in order to obtain the Mathieu amalgam. In the first step we *twist* the subamalgam $\{H_1, H_2\}$, preserving its type. By Goldschmidt's lemma in Section 2.2 this can be achieved by premultiplying the embedding of H_{12} in H_2 by an automorphism α of H_{12}. This automorphism should not be realizable in the normalizers in $\mathrm{Aut}(H_1)$ and $\mathrm{Aut}(H_2)$ of the images of H_{12} in H_1 and H_2, respectively. In particular, α must be an outer automorphism of H_{12}. First we describe the automorphism group of H_{12} in general terms and then make a specific choice for α that will be convenient for the subsequent constructions.

Let $R_2^{(4)}$, $R_2^{(5)}$ and $R_2^{(4\frac{1}{2})}$ denote the subgroups in R_2 generated by the transvections τ_{4i}, τ_{5i} and $\tau_{4i}\tau_{5i}$ taken for $i = 1, 2, 3$. Let us use a fresh name L for the Levi $L_3(2)$-subgroup K_2^+. Then each of the $R_2^{(j)}$s is a natural L-module, any two of which will generate the whole of R_2, which is the direct sum of two copies of the natural L-module. Furthermore, R_2L is the index-2 commutator subgroup H_{12}' of H_{12}. Let us have a look at $\mathrm{Aut}(H_{12}')$ first. Since the centre of H_{12}' is manifestly trivial, the inner automorphism group can be identified with H_{12}'. An outer automorphism might permute the three copies of the natural L-module in R_2 and also permute the classes of $L_3(2)$-complements to R_2. By the *double complement* Lemma 3.10 there are four classes of such complements. In fact, $R_2^{(j)}L$ contains two classes of complements: L represents one of them and a representative of the other one we denote by $L^{(j)}$, so that

$$R_2^{(j)}L = R_2^{(j)}L^{(j)} \text{ for } j = 4,\ 5,\ 4\tfrac{1}{2}.$$

We will use calligraphic letters to denote the H_{12}'-conjugacy class containing a given representative, so that

$$\mathcal{C} = \{\mathcal{L},\ \mathcal{L}^{(4)},\ \mathcal{L}^{(5)},\ \mathcal{L}^{(4\frac{1}{2})}\}$$

is the complete set of classes of $L_3(2)$-complements to R_2 in H_{12}'.

Lemma 4.3 *Out*$(H_{12}') \cong \mathrm{Sym}(\mathcal{C}) \cong S_4$.

Proof We need to show that $\mathrm{Out}(H_{12}')$ induces on the set $\mathcal{C}$ its full symmetric group S_4. First of all the automorphism group is transitive on the complements, since R_2 is abelian and any complement is suitable to build up H'_{12} as a semidirect product. Furthermore, H'_{12} is normal in H_2 and the latter group, while preserving $\mathcal{L}$, induces $S_3 \cong K_2^-$ on $\{R_2^{(4)}, R_2^{(5)}, R_2^{(4\frac{1}{2})}\}$. Therefore H_2 induces S_3 on the classes in $\mathcal{C}$ other than $\mathcal{L}$ and the claim follows. On the other hand, $\mathrm{Out}(H'_{12})$ is faithful on $\mathcal{C}$, since an automorphism which stabilizes every class of complements must also stabilize every copy of the natural L-module in R_2 and hence it must be inner on $R_2^{(j)}L$ for every j in our range and thus must be inner on the whole of H'_{12}. □

The above lemma is an important achievement, but what we need is the automorphism group of H_{12}. The centre of H_{12} is trivial, H_{12} contains H'_{12} as a characteristic subgroup, and $H_{12} = \langle H'_{12}, \tau_{54}\rangle$. It is rather clear that only the identity automorphism of H_{12} centralizes H'_{12}. One way of seeing this is to observe that τ_{54} is the only involution in H_{12} which commutes with an element of order 7 from $L = K_2^+$. Therefore $\mathrm{Aut}(H_{12})$ is the normalizer of H_{12} in $\mathrm{Aut}(H'_{12})$. The transvection τ_{54} centralizes $R_2^{(5)}$ and L. Since it permutes $R_2^{(4)}$ and $R_2^{(4\frac{1}{2})}$ it must also permute $\mathcal{L}^{(4)}$ and $\mathcal{L}^{(4\frac{1}{2})}$, stabilizing $\mathcal{L}^{(5)}$. This enables us to conclude that the image of H_{12} in $\mathrm{Out}(H'_{12})$ is the subgroup of order 2 generated by the transposition

$$\widehat{\tau_{54}} = (\mathcal{L})(\mathcal{L}^{(5)})(\mathcal{L}^{(4)}\,\mathcal{L}^{(4\frac{1}{2})})$$

(which is the image of τ_{54}). By Lemma 4.3 we have the following.

Lemma 4.4 *The group* $\mathrm{Out}(H_{12})$ *has order* 2 *and it is generated by the coset of an outer automorphism* α*, whose image in* $\mathrm{Out}(H'_{12}) \cong S_4$ *is the permutation*

$$\widehat{\alpha} = (\mathcal{L}\,\mathcal{L}^{(5)})(\mathcal{L}^{(4)}\,\mathcal{L}^{(4\frac{1}{2})}).$$

□

4.5 Building up $\{G_1, G_2\}$

Our construction of the Mathieu amalgam starts with a rank-2 amalgam $\{G_1, G_2\}$ which is the subamalgam $\{H_1, H_2\}$ of the $L_5(2)$-amalgam, twisted (in the sense of Goldschmidt's lemma) by the automorphism α of H_{12} defined in the previous section. By Lemma 4.4 the automorphism α represents the only non-identity coset of $\mathrm{Inn}(H_{12}) \cong H_{12}$ in $\mathrm{Aut}(H_{12})$. The amalgams $\{H_1, H_2\}$ and $\{G_1, G_2\}$ have the same type, although they are not isomorphic. The non-isomorphism will become crystal clear when G_3 is adjoined, but it already follows from Goldschmidt's lemma together with the following well-known result whose proof we include for completeness since it also illuminates the special behaviour of $L_3(2)$ among the linear groups over $GF(2)$.

Lemma 4.5 *The group* $H_1 \cong 2^4 : L_4(2)$ *is complete in the sense that* $Aut(H_1) = Inn(H_1) \cong H_1$.

Proof Since the centre of H_1 is trivial, all we have to prove is that every $\beta \in \mathrm{Aut}(H_1)$ is inner. Since R_1 is characteristic in H_1 and since $H_1/R_1 \cong L_4(2)$ is the automorphism group of R_1, we can adjust β by an inner automorphism to make it act trivially on R_1. Since the outer automorphism group of H_1/R_1 is of order 2 generated by the contragredient automorphism which sends the natural module R_1 onto its dual, the adjusted β acts trivially on H_1/R_1. Let $K_1 \cong L_4(2)$ be the Levi complement to R_1 in H_1. To prove the lemma it is sufficient to show that β normalizes K_1. Let T_9 be a Sylow 3-subgroup in K_1 and T_3 be a subgroup of order 3 in T_9 which acts on R_1 fixed-point freely. Notice that in the A_8-incarnation of K_1 the subgroup T_3 is generated by a 3-cycle while T_9 is generated by two disjoint 3-cycles. Then, because of the fixed-point freeness of the actions on R_1, we have

$$N_{H_1}(T_9) = N_{K_1}(T_9) \cong 3^2 : D_8,$$
$$N_{H_1}(T_3) = N_{K_1}(T_3) \cong (S_3 \times S_5)^+.$$

By adjusting β further by an inner automorphism induced by an element from R_1 (since R_1 is abelian this does not spoil the previous adjustment) we make β such that it will centralize T_9 and hence T_3 as well. Then β normalizes the normalizers of T_9 and T_3 in H_1. By the above these two normalizers are contained in K_1 and they clearly generate the whole of K_1. Therefore $\beta(K_1) = K_1$ as required. Since β has been adjusted to act trivially both on R_1 and on $H_1/R_1 \cong K_1$, it must be the identity automorphism. □

The amalgams $\{H_1, H_2\}$ and $\{G_1, G_2\}$ can both be seen as subamalgams in $\{H_1, \mathrm{Aut}(H_{12})\}$, where G_1 is identified with H_1, while H_2 and G_2 are the preimages in $\mathrm{Aut}(H_{12})$ of the two S_3-subgroups in $\mathrm{Out}(H_{12}) \cong \mathrm{Sym}(\mathcal{C}) \cong S_4$ which are the stabilizers of $\mathcal{L}$ and $\mathcal{L}^{(5)}$, respectively.

In order to smoothen the adjoining of G_3 to the amalgam $\{G_1, G_2\}$ we have just constructed, we make an explicit choice for α so that it normalizes a Sylow 3-subgroup of K_3^+. By the *transvection product rule* Lemma 4.1 the element $t = \tau_{12}\tau_{21}$ has order 3 and hence generates a Sylow 3-subgroup in K_3^+ and also in L. Let $S \cong F_{21}$ be the normalizer of a Sylow 7-subgroup in L, which contains t. Notice that there are two choices for S which are equally good for our purpose. Then, by virtue of the maximal subgroup structure of L, the latter is generated by S together with an involution from L, which inverts t, and τ_{21} is such an involution, so that

$$L = \langle S, \tau_{21} \rangle.$$

By the *product* principle in Section 3.10, in order to obtain another $L_3(2)$-complement in H'_{12} containing S we have to substitute τ_{21} by some other involution inverting t. Since R_2 is the direct sum of two copies of the natural L-module and t centralizes a single non-zero vector in each copy, we obtain three further complements:

$$L^{(j)} = \langle S, \tau_{21}\tau_{j3} \rangle, \quad j = 4,\ 5,\ 4\tfrac{1}{2},$$

where $\tau_{4\frac{1}{2}3} = \tau_{43}\tau_{53}$.

In view of the discussions in Section 3.10, the mapping which centralizes $R_1 R_2 S = O_2(H_{12})S$ and performs the permutation

$$(\tau_{21}, \tau_{21}\tau_{53})(\tau_{21}\tau_{43}, \tau_{21}\tau_{43}\tau_{53})$$

extends uniquely to an outer automorphism α of H_{12}, whose image in $\mathrm{Out}(H'_{12})$ acts on the set $\mathcal{C}$ of classes of $L_3(2)$-complements exactly as the above-introduced permutation $\widehat{\alpha}$ does. This is the automorphism α we will stick with. Notice that inside $\{H_1, \mathrm{Aut}(H'_{12})\}$ we have

$$\{G_1, G_2\} = \{H_1, \alpha(H_2)\}.$$

4.6 An Explicit Form of α

For further calculation it appears convenient to find an explicit form of the automorphism α. In order to achieve this we take S (which is a Sylow 7-subgroup of $L = K_2^+ \cong L_3(2)$) to be generated by the element s presented by the following matrix:

$$s = \begin{pmatrix} 0 & 1 & 1 \\ 1 & 0 & 0 \\ 0 & 1 & 0 \end{pmatrix}.$$

We also view s as an element of $H \cong L_5(2)$, where the above matrix is in the top left corner and is completed by the identity matrix at the 2×2 bottom right corner (which is the identity element of K_2^-). In these terms $t = \tau_{12}\tau_{21}$ is represented by the matrix

$$t = \begin{pmatrix} 0 & 1 & 0 \\ 1 & 1 & 0 \\ 0 & 0 & 1 \end{pmatrix},$$

and one easily checks that $t^{-1}st = s^4$. As above

$$R_2^{(5)} = \langle \tau_{51}, \tau_{52}, \tau_{53} \rangle,$$

so that $R_2^{(5)} L = R_2^{(5)} L^{(5)}$ and α is an automorphism of that group, which centralizes

$$R_2^{(5)} : \langle s, t \rangle \cong 2^3 : F_{21}$$

and swaps L with $L^{(5)}$. We have seen already that

$$\alpha \colon \tau_{21} \mapsto \tau_{21}\tau_{53}.$$

This rule can be extended to all the involutions of L. In fact, τ_{21} normalizes $\langle t \rangle$ and τ_{53} is the only involution in $R_2^{(5)}$ which is centralized by t. If we put

$$t_i := (s^i)^{-1} t s^i \quad \text{for } 0 \le i \le 6$$

then we obtain seven generators of the Sylow 3-subgroups of $\langle s, t \rangle \cong F_{21}$, where $t_0 = t$. Inside L the order-3 subgroup $\langle t_i \rangle$ has normalizer isomorphic to S_3 and hence it is normalized (inverted) by three involutions. It is clear that an involution cannot invert two different t_is, since otherwise it would normalize the subgroup of order 7 in S, which they generate. In this way we have accounted for all of the $21 = 7 \times 3$ involutions in L and in order to find the image of an involution $\tau \in L$ under α we should carry out the following steps:

(i) find the unique i in the range $0 \le i \le 6$ such that τ inverts t_i;
(ii) find the unique involution $\sigma \in R_2^{(5)}$ centralized by t_i;
(iii) set $\alpha \colon \tau \mapsto \tau\sigma$.

By applying this rule we find

$$\alpha \colon \tau_{31} \mapsto \tau_{31}\tau_{52}, \quad \alpha \colon \tau_{32} \mapsto \tau_{32}\tau_{51}\tau_{52}, \quad \alpha \colon \tau_{23} \mapsto \tau_{23}\tau_{51}\tau_{53}$$

and can easily extend this to further involutions in L if necessary. Overall this suggests the following matrix arrangement of the generators of G_2, where in the ij-entry we place the image of τ_{ij} under α:

$$G_2 = \begin{pmatrix} 1 & \tau_{12}\tau_{53} & \tau_{13}\tau_{52}\tau_{53} & 0 & 0 \\ \tau_{21}\tau_{53} & 1 & \tau_{23}\tau_{51}\tau_{53} & 0 & 0 \\ \tau_{31}\tau_{52} & \tau_{32}\tau_{51}\tau_{52} & 1 & 0 & 0 \\ \tau_{41} & \tau_{42} & \tau_{43} & 1 & \tau_{45}^{\alpha} \\ \tau_{51} & \tau_{52} & \tau_{53} & \tau_{54} & 1 \end{pmatrix}.$$

In the above matrix the top left (3×3)-block contains the generators of $L^{(5)}$ which commute with

$$\langle \tau_{45}^{\alpha}, \tau_{54} \rangle \cong S_3.$$

Notice that α normalizes $R_3 = \langle \tau_{31}, \tau_{32}, \tau_{41}, \tau_{42}, \tau_{51}, \tau_{52} \rangle$.

4.7 Incorporating G_3

In order to extend $\{G_1, G_2\}$ to the Mathieu amalgam, we consider the subamalgam

$$\mathcal{A}_3 = \{N_{G_1}(R_3), N_{G_2}(R_3)\},$$

and identify its core. The subgroup R_3 is in the core, since $\mathcal{A}_3$ is formed by the normalizers of R_3 in G_1 and G_2. The core of

$$\mathcal{H}_3 := \{N_{H_1}(R_3), N_{H_2}(R_3)\}$$

is $R_3 K_3^+$, where $K_3^+ = \langle t \rangle : \langle \tau_{21} \rangle$ with $t = \tau_{12}\tau_{21}$. On the other hand,

$$N_{G_1}(R_3) = N_{H_1}(R_3) \quad \text{and} \quad N_{G_2}(R_3) = \alpha(N_{H_2}(R_3)).$$

Since α centralizes t, the latter is in the core, but $\alpha(\tau_{21}) = \tau_{21}\tau_{53}$ is not in K_3^+, therefore τ_{21} is not in the core. Thus we have shown that $T_3 = R_3\langle t \rangle$ is a normal subgroup of the core of $\mathcal{A}_3$, and the subsequent constructions will demonstrate that T_3 is the whole of the core.

Lemma 4.6 *$Aut(T_3) \cong 2^6 : \Gamma L_3(4)$ and* $\mathrm{Out}(T_3) \cong P\Gamma L_3(4)$.

Proof Notice that T_3 is a normal extension of the subgroup R_3, which is elementary abelian of order 2^6, by an order-3 subgroup $\langle t \rangle$ generated by the element t, which acts on R_3 fixed-point freely. This implies that T_3 is centreless and that the automorphism group of T_3 preserves the $GF(4)$-vector space structure on R_3 induced by the action of t. □

Lemma 4.7 *The subgroups R_3 is self-centralized in $N_{G_1}(R_3)$ and in $N_{G_2}(R_3)$.*

Proof The amalgam $\mathcal{H} = \{H_1, H_2, H_3\}$ is clearly constrained with respect to R_3 and therefore R_3 is self-centralized in $N_{H_1}(R_3)$ and in $N_{H_2}(R_3)$. On the other hand, $G_1 = H_1$, while $G_2 = \alpha(H_2)$. It is easily seen from the matrix arrangement of the generators of G_2 given in the previous section that R_3 is normalized by α (this is why α has been taken as it is). Hence the claim follows. □

By the above lemma $\mathcal{A}_3$ can be viewed as a subamalgam in $\mathrm{Aut}(T_3)$, and we define G_3 to be the subgroup in $\mathrm{Aut}(T_3)$ generated by this subamalgam. Then $G_{i3} = N_{G_i}(R_3)$ for $i = 1, 2$, and in order to accomplish the construction of the *Mathieu amalgam*

$$\mathcal{A} = \{G_1, G_2, G_3\}$$

it remains only to identify the isomorphism type of G_3. Notice that $\mathcal{A}$ is constrained with respect to T_3 by its definition as a subamalgam in $\{G_1, G_2, \mathrm{Aut}(R_3)\}$.

In order to identify G_3, let us have a closer look at $\mathcal{H}_3$ and its members. Let $\Xi = (\mathcal{P}_4, \mathcal{L}_4)$ be the projective plane of order 4 associated with the $GF(4)$-vector space structure on R_3. Then $\mathcal{P}_4$ is the set of orbits of $\langle t \rangle$ on R_3, while $\mathcal{L}_4$ consists of the subgroups of order 2^4 in R_3 generated by pairs of such orbits and T_3 is the kernel of the action of H_3 on Ξ. Let $\Phi(\tau_{21})$ be the Fano subplane in Ξ formed by the points and lines fixed by τ_{21}. Then we have seen that the image of H_3 in $P\Gamma L_3(4) = \mathrm{Aut}(\Xi)$ is precisely the stabilizer of $\Phi(\tau_{21})$. Furthermore, H_{13} is the stabilizer of the point

$$p = \{\tau_{51}, \tau_{52}, \tau_{51}\tau_{52}\},$$

while H_{23} is the stabilizer of the line

$$l = \langle \tau_{41}, \tau_{42}, \tau_{51}, \tau_{52} \rangle.$$

The action of G_3 on Ξ is generated by the actions of G_{13} and of G_{23}. The action of G_{13} coincides with that of H_{13} and, by virtue of the above, it is the stabilizer in $\mathrm{Aut}(\Xi)$ of the Fano subplane $\Phi(\tau_{21})$ together with the point p in this subplane. The action of G_{23} is the α-twist of that of H_{23}, thus G_{23} is the stabilizer of the Fano subplane $\alpha(\Phi(\tau_{21})) = \Phi(\tau_{21}\tau_{53})$ together with the line $\alpha(l) = l$ in that subplane. The following lemma involves the *transfer principle* Lemma 3.20.

Lemma 4.8 *The subgroup G_3 is the stabilizer of the hyperoval*

$$\Theta := \Theta(\Phi(\tau_{21}), p).$$

Proof The element τ_{53} acts on $\Phi(\tau_{21})$ as the transvection with centre p and axis l, therefore $\Phi(\tau_{21}\tau_{53})$ is the reflection of $\Phi(\tau_{21})$ with respect to the flag (p, l) in the sense of the *flag reflection principle* Lemma 3.19. □

Lemma 4.8 is illustrated by Figure 4.1, where the reflection is with respect to the flag formed by the horizontal line and the point in the centre of it. The plane $\Phi(\tau_{21})$ is formed by the four points above the line and the three points to the

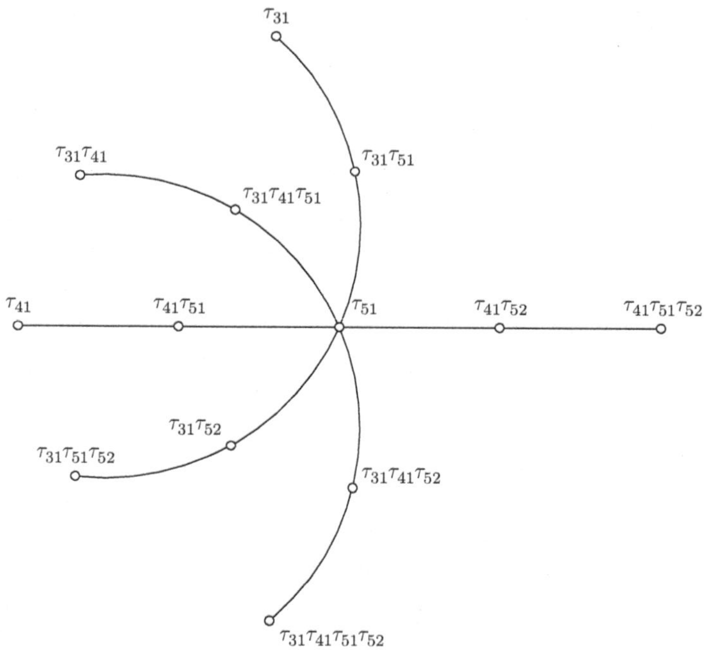

Figure 4.1

left on the horizontal line, while $\Phi(\tau_{21}\tau_{53})$ is formed by the same three points on the line and the four points below it.

Thus $G_3 \cong 2^6 : 3 \cdot \mathrm{Sym}_6$ is the stabilizer in $\mathrm{Aut}(T_3) \cong 2^6 : \Gamma L_3(4)$ of the hyperoval Θ inside the projective plane Ξ of order 4. Furthermore, G_{13} is the stabilizer of the partition of Θ into three disjoint pairs (of hooves of the lines in $\Phi(\tau_{21})$ passing through p), while G_{23} is the stabilizer of the pair of hooves of the line l with respect to either $\Phi(\tau_{21})$ or $\Phi(\tau_{21}\tau_{53})$. Thus the construction of the Mathieu amalgam

$$\mathcal{A} = \{2^4 : L_4(2),\ 2^6 : (L_3(2) \times S_3),\ 2^6 : 3 \cdot \mathrm{Sym}_6\}$$

is complete and the amalgam is characterized by the following direct consequence of the construction.

Theorem 4.9 *The Mathieu amalgam is the unique amalgam which is constrained with respect to R_3 and whose first two members constitute an amalgam having the same type as $\{H_1, H_2\}$ but not isomorphic to it.* □

4.8 The Minimality of G_3

In this section we discuss the special feature of the Mathieu amalgam in terms of the certain *minimality* of the member G_3. Let

$$\varphi: G_1 \cup G_2 \to F$$

be a faithful completion of the rank-2 amalgam $\{G_1, G_2\}$. This means that the restriction $\varphi^{(i)}$ of φ to G_i is a monomorphism and $\varphi^{(1)}(g) = \varphi^{(2)}(g)$ for every $g \in G_{12}$. Then we can define

$$G_3^{(\varphi)} = \langle \varphi^{(1)}(G_{13}), \varphi^{(2)}(G_{23}) \rangle.$$

Then $G_3^{(\varphi)}$ is a completion of the subamalgam $\mathcal{X}_3 = \{G_{13}, G_{23}\}$ whose core is $T_3 \cong 2^6 : 3$. Put

$$C_3^{(\varphi)} = C_{G_3^{(\varphi)}}(\varphi(T_3)).$$

Then we have proved the following *minimality* principle.

Lemma 4.10 *A faithful completion φ of the rank 2 amalgam $\{G_1, G_2\}$ is a faithful completion of the Mathieu amalgam if and only if $C_3^{(\varphi)} = 1$. Any completion group $G_3^{(\varphi)}$ of the amalgam $\mathcal{X}_3$ possesses a homomorphism onto G_3 with kernel $C_3^{(\varphi)}$.* □

By Lemma 4.9 we can redefine the Mathieu amalgam as follows:

$$\mathcal{A} = \{G_1, G_2, U_3/C_{X_3}(T_3)\},$$

where X_3 is the universal completion of the amalgam $\mathcal{X}_3$ and T_3 is identified with its image in X_3

Notice that $G_3^{(\varphi)}/\varphi(T_3)C_3^{(\varphi)} \cong S_6$ and the images of G_{13} and G_{23} in this quotient constitute a pair of subgroups, each isomorphic to $S_4 \times 2$, which intersect in a Sylow 2-subgroup of order 16. The coset graph of the completion

$$\{S_4 \times 2, S_4 \times 2\} \to S_6$$

is the incidence graph Θ of a quad. The graph Θ has 2-subsets of a six-element set and the partitions of this 6-set into three pairs as vertices with the natural adjacency relation (of valency 3). Therefore, if φ is the universal completion, then $C_3^{(\varphi)}$ is a free group whose generators are indexed by the fundamental cycles of Θ. The number of fundamental cycles is the number of edges minus the number of vertices plus one. In the case considered here it is

$$45 - 30 + 1 = 16,$$

which is the dimension of the Steinberg module of $S_6 \cong Sp_4(2)$. We summarize this section by the following lemma.

Lemma 4.11 *Let $\mathcal{X}_3 = \{G_{13}, G_{23}\}$ be a subamalgam in $\{G_1, G_2\}$ formed by the normalizers of T_3, let*

$$\varphi \colon \mathcal{X}_3 \to X_3$$

be the universal completion of $\mathcal{X}_3$, and let $C_3^{\varphi} = C_{X_3}(\varphi(T_3))$. Then

(i) $X_3/\varphi(T_3)C_3^{\varphi} \cong S_6 \cong Sp_4(2)$;
(ii) *the largest abelian quotient of C_3^{φ} is isomorphic to the* 16*-dimensional Steinberg module of the quotient group in* (i). □

Finally, if φ as above is the universal completion of the rank-2 amalgam with the completion group F then the universal completion group of the Mathieu amalgam is the quotient of F over the normal closure of $C_3^{(\varphi)}$. Whether or not this completion is faithful is potentially a highly non-trivial question.

5

M_{24} As Universal Completion

In this chapter we determine the universal completion of the Mathieu amalgam to construct the Mathieu group M_{24} in its new incarnation.

5.1 Completing the Mathieu Amalgam

According to the orthodox paradigm of the amalgam method, after constructing the Mathieu amalgam

$$\mathcal{A} = \{G_1, G_2, G_3\}$$

we have to decide whether its universal completion is faithful. To answer this question affirmatively it is sufficient to construct a faithful completion. For the Mathieu amalgam it proves easiest to find such a completion inside the symmetric group of degree 24. When one is dealing with permutation groups, it is helpful to keep in mind the following well-known *centralizer principle.*

Lemma 5.1 *A permutation group whose element stabilizers are pairwise different has a trivial centralizer in the corresponding symmetric group.* □

Let $\mathcal{P}$ be a 24-set. We are going to construct an injection

$$\varphi\colon G_1 \cup G_2 \cup G_3 \to \mathrm{Sym}(\mathcal{P}) \cong S_{24}$$

such that the restriction of φ to each member of the amalgam is a monomorphism.

First make G_1 act on $\mathcal{P}$ with two orbits $\mathcal{P}_8$ and $\mathcal{P}_{16}$ of sizes 8 and 16, respectively. On $\mathcal{P}_8$ the action is through the surjective homomorphism

$$G_1 \to \mathrm{Alt}(\mathcal{P}_8) \cong A_8 \cong L_4(2) \cong G_1/R_1$$

with kernel $R_1 \cong 2^4$. The action on $\mathcal{P}_{16}$ is on the cosets of K_1, which is an $L_4(2)$-complement to R_1. The action on $\mathcal{P}_{16}$ possesses the following transparent geometrical description. The group G_1 is the stabilizer in $L_5(2)$ of the 1-space in $V_5(2)$ spanned by b_5. Therefore G_1 is also the stabilizer of the hyperplane b_5^*. Then the set $\mathcal{P}_{16}$ can be viewed as the complement of b_5^*, so that $\mathcal{P}_{16}$ becomes the set of vectors in $V_5(2)$ with non-zero last coordinate in the basis $(b_1, \ldots, b_5)$ and the vector stabilized by K_1 is b_5 itself.

Having constructed the action of G_1 on $\mathcal{P}$, we restrict it to G'_{12}. On the orbit $\mathcal{P}_8$ the subgroup G'_{12} induces the maximal parabolic $2^3 : L_3(2)$ of K_1, where R_2 induces the regular normal subgroup of order 2^3 with kernel $V^{(5)} = R_1 \cap R_2$. The set $\mathcal{P}_8$, when viewed as a G'_{12}-set, will be decorated by the superscript (5). Under the action of G'_{12} the set $\mathcal{P}_{16}$ splits into two orbits of size 8 each. The G'_{12}-orbit of a vector from $\mathcal{P}_{16}$ is determined by the value of its b_4-coordinate. The orbits of vectors with zero and non-zero b_4-coordinates will be denoted by $\mathcal{P}_8^{(4)}$ and $\mathcal{P}_8^{(4\frac{1}{2})}$, respectively. Notice that the former orbit is contained in the hyperplane b_4^*, while the latter is in the hyperplane $b_4^* + b_5^*$.

In total there are three G'_{12}-orbits on $\mathcal{P}$ of size 8 each, and they are listed in the leftmost column of Table 5.1. It follows from the above paragraph that the kernels at these orbits are natural $L_3(2)$-submodules in R_2 and they are specified in the second column. Given an $L_3(2)$-complement $L^{(\alpha)}$ to R_2 in G'_{12} and an orbit $\mathcal{P}_8^{(\beta)}$, the action of $L^{(\alpha)}$ on $\mathcal{P}_8^{(\beta)}$ is either transitive, or with two orbits with lengths 1 and 7. These actions are specified in the four columns to the right in Table 5.1. The entries in these columns can be justified by making use of three principles presented in the following lemma.

Lemma 5.2 *The following assertions hold:*

(i) *L acts transitively on $\mathcal{P}_8^{(5)}$;*

(ii) *L fixes $b_5 \in \mathcal{P}_8^{(4)}$ and $b_4 + b_5 \in \mathcal{P}_8^{(4\frac{1}{2})}$;*

(iii) *$L^{(\alpha)}$ and $L^{(\gamma)}$ act similarly on $\mathcal{P}_8^{(\beta)}$ if and only if $R_2^{(\beta)} L^{(\alpha)} = R_2^{(\beta)} L^{(\gamma)}$.*

Proof The assertion (i) immediately follows from Table 3.1. Since L is a Levi complement in $L_5(2)$ with respect to the decomposition $V_5(2) = \langle b_1, b_2, b_3 \rangle \oplus \langle b_4, b_5 \rangle$, (ii) follows. Finally (iii) is immediately evident since R_2^{β} is the kernel of G'_{12} at $\mathcal{P}_8^{(\beta)}$. □

Table 5.1 indicates the stabilizers of elements in every orbit and thus uniquely determines $\varphi(G'_{12})$. We construct $\varphi(G_2)$ to satisfy the equality

$$N_{\mathrm{Sym}(\mathcal{P})}(\varphi(G'_{12})) = \varphi(G_2).$$

Table 5.1

Orbit	Kernel	L	$L^{(5)}$	$L^{(4)}$	$L^{(4\frac{1}{2})}$
$\mathcal{P}_8^{(5)}$	$R_2^{(5)}$	8	8	$1+7$	$1+7$
$\mathcal{P}_8^{(4)}$	$R_2^{(4)}$	$1+7$	8	$1+7$	8
$\mathcal{P}_8^{(4\frac{1}{2})}$	$R_2^{(4\frac{1}{2})}$	$1+7$	8	8	$1+7$

In order to achieve this we define $\varphi(G_2)$ to be the action of $G^{(2)}$ on the cosets of $R_2^{(4)}L$. Since $\mathrm{Out}(G'_{12})$ is the symmetric group of the four classes of $L_3(2)$-complements to R_2 in G'_{12}, while G_2 is the stabilizer in $\mathrm{Aut}(G'_{12})$ of the class containing $L^{(5)}$, we conclude that the G_2-class containing $R_2^{(4)}L$ splits into three G'_{12}-classes with representatives $R_2^{(4)}L$, $R_2^{(5)}L^{(4)}$ and $R_2^{(4\frac{1}{2})}L^{(4\frac{1}{2})}$. Thus the restrictions to G'_{12} of $\varphi(G_1)$ and $\varphi(G_2)$ are indeed similar.

By the construction $\varphi(G'_{12})$ is normal in $\varphi(G_2)$. In view of the *centralizer* principle Lemma 5.1, it is immediately evident from Table 5.1 that the centralizer of $\varphi(G'_{12})$ in $\mathrm{Sym}(\mathcal{P})$ is trivial. Therefore $\varphi(G_2)$ is the full normalizer of $\varphi(G'_{12})$ in $\mathrm{Sym}(\mathcal{P})$. Since G'_{12} is normal in G_{12}, the action of the latter is uniquely determined irrespective of whether it is restricted from G_1 or from G_2. In fact, $\varphi(G_{12})$ is the stabilizer in $\varphi(G_2)$ of the class of L and the following lemma holds.

Lemma 5.3 *The constructed mapping*

$$\varphi\colon G_1 \cup G_2 \to \mathrm{Sym}(\mathcal{P})$$

is a faithful completion of the subamalgam in the Mathieu amalgam formed by the first two members. □

In order to extend φ in Lemma 5.3 to the whole of the Mathieu amalgam we put

$$G_3^{(\varphi)} = \langle \varphi(G_{13}), \varphi(G_{23}) \rangle_{\mathrm{Sym}(\mathcal{P})}$$

and demonstrate that $\varphi(T_3)$ has a trivial centralizer in $\mathrm{Sym}(\mathcal{P})$. Since that centralizer contains $C_3^{(\varphi)}$, by the *minimality* principle Lemma 4.10 this will provide us with an isomorphism between $G_3^{(\varphi)}$ and G_3. To this end all we need is the *centralizer principle* Lemma 5.1 together with the following *A_4-principle*.

Lemma 5.4 *The subgroup* $\varphi(T_3) \cong 2^6 : 3$ *acts on* $\mathcal{P}$ *with six orbits, each of length 4. The induced action on each orbit is the alternating group* A_4 *of degree 4, and the kernels at different orbits are different.*

Proof Let us consider G_{13}/R_1. In the A_8-incarnation of G_1/R_1 we observe that G_{13} is the stabilizer of a partition of $\mathcal{P}_8$ into two subsets of size 4 each, while in the $L_4(2)$-incarnation it is the stabilizer of a 2-subspace in the natural four-dimensional $GF(2)$-module b_5^*. We claim that R_3 acting on $\mathcal{P}$ has six orbits, each of length 4. Inside $\mathcal{P}_8$ these orbits are the members of the $8 = 4{+}4$ partition, while, given a vector $v \in \mathcal{P}_{16}$, its R_3-orbit is determined by the values of the b_3- and b_4-coordinates (with all four possibilities allowed) and the claim follows. □

Let S be the set of R_3-orbits on $\mathcal{P}$. Then knowledge of $\varphi(G_1)$ and $\varphi(G_2)$ is sufficient to demonstrate that the kernels at different orbits are different T_3-invariant subgroups of R_3. Furthermore, $\varphi(G_{13})$ preserves S and stabilizes a pair of orbits in $\mathcal{P}_8$, while the subgroup $\varphi(G_{23})$ also preserves S and stabilizes a partition of S, into three pairs. Therefore G_3^{φ} preserves S, inducing on it the full symmetric group $\mathrm{Sym}(S) \cong S_6$, with T_3 being in the kernel. Thus G_3^{φ} acts on S via its factor group

$$G_3^{(\varphi)}/\varphi(T_3)C_3^{(\varphi)} \cong S_6.$$

Therefore, every R_3-orbit is also a T_3-orbit since by the A_4-principle Lemma 5.4 an element of order 3 in T_3 acts on R_3 fixed-point freely. Thus there is no space to allocate a non-trivial centralizer C_3^{φ}. Hence by Lemma 5.3 we have proved the following theorem.

Theorem 5.5 *The isomorphism* $G_3^{(\varphi)} \cong G_3$ *holds and* φ *is a faithful completion of the Mathieu amalgam into* $\mathrm{Sym}(\mathcal{P}) \cong S_{24}$. □

It should be clear that $\mathcal{P}_8$ is an octad, $\mathcal{P} = \mathcal{P}_8^{(5)} \cup \mathcal{P}_8^{(4)} \cup \mathcal{P}_8^{(4\frac{1}{2})}$ is a trio and the partition S of $\mathcal{P}$ into six R_3-orbits is a sextet.

5.2 The Octad Graph As the Coset Graph

Let G be a faithful completion of the Mathieu amalgam $\mathcal{A} = \{G_1, G_2, G_3\}$. By Theorem 5.5, G exists, and the goal of this section is to determine the order of G. This will be achieved through the reconstruction of the octad graph Γ as a graph on the set of right cosets of G_1 in G, where two coset-vertices are adjacent whenever they intersect a common coset of G_2. Thus we put

$$V(\Gamma) = \{G_1h \mid h \in G\}$$

and let $v_1 = G_1 \in V(\Gamma)$ be the coset-vertex containing the identity element. The group G acts naturally on $V(\Gamma)$ via

$$g : G_1h \mapsto G_1hg.$$

With respect to this action, G_1 is the stabilizer of v_1 and the stabilizer of a generic coset-vertex G_1h is the conjugate $G_1^h := h^{-1}G_1h$ of G_1 by a representative h of the coset. As declared, two coset-vertices are adjacent in the graph Γ whenever they intersect a common coset of G_2. This definition admits the following specification. Consider the orbit T of v_1 under G_2. Then

$$T = \{G_1h \mid h \in G_2\}, \quad |T| = [G_1 : G_{12}] = 3,$$

and by definition T is a complete subgraph in Γ. Thus T is a triangle and the edges of Γ are the images under G of the three edges in T. In what follows we adopt the notation

$$R_{ij} = R_i \cap R_j, \quad \text{where } 1 \le i < j \le 3.$$

Since G_2 induces on T the natural action of S_3, it is clear that G'_{12} is the vertex-wise stabilizer of T in G and that R_1 induces on T an action of order 2 (permuting two vertices in $T \setminus \{v_1\}$) with kernel R_{12}, which is a hyperplane in R_1. Since $G_{12} = N_{G_1}(R_{12})$, since G_1 permutes transitively the hyperplanes in R_1, and since Γ is clearly not a complete graph, we have the following *first neighbourhood* description. Here and elsewhere $\Gamma_i(x)$ denotes the set of vertices in Γ, which are at distance i from x in Γ.

Lemma 5.6 *The set $\Gamma_1(v_1)$ consists of* 30 *vertices. R_1 acts on this set with* 15 *orbits of length* 2 *each, indexed in the natural way by the hyperplanes in R_1. Every R_1-orbit together with v_1 is a triangle, which is the image of T under an element of G_1. Two vertices in $\Gamma_1(v_1)$ are adjacent if and only if they are in the same R_1-orbit, so that in distance-regular terms $k = 30$, $a_1 = 1$ and $b_1 = 28$. The subgroup G_1 acts transitively on the pairs of non-adjacent vertices in $\Gamma_1(v_1)$ and hence it acts transitively on $\Gamma_2(v_1)$.*

Proof The group G_1 acts on $\Gamma_1(v_1)$ as on the cosets of its subgroup G'_{12}. Viewing $G_1(v_1)$ as the stabilizer of the vector b_5 in $L_5(2)$, we can identify v_1 with b_5 and $\Gamma_1(v_1)$ with $V_5(2) \setminus \langle b_5 \rangle$, preserving the natural action. Two vector-vertices in $\Gamma_1(v_1)$ are adjacent whenever their sum is b_5, so that $T = \{b_5, b_4, b_5 + b_4\}$. Since $\{G_1, G_2\}$ is not contained in $L_5(2)$, this identification is *local* in the sense that it does not go beyond $\Gamma_1(v_1)$. □

Next we prove the following *second neighbourhood* description, where a quad is a 15-vertex subgraph whose vertices are indexed by 2-subsets of a 6-set with two subsets adjacent whenever they are disjoint.

Lemma 5.7 *Let S be the subgraph induced by the images of v_1 under G_3. Then S is a quad, there are $35 = [G_1 : G_{13}]$ quads containing v_1 indexed by the 2-subspaces in R_1, which are kernels at the quads. Every vertex from $\Gamma_2(v_1)$ is in a unique such quad, so that $|\Gamma_2(v_1)| = 35 \cdot 8 = 280$. The subgroup R_1 acts on $\Gamma_2(v_1)$ with orbits of length 4, with two orbits have the same kernel only when they are in the same quad through v_1. The quad S stabilized by G_3 corresponds to R_{13}. If $S(U)$ is the quad containing v_1 fixed by a 2-subspace U in R_1, and if $T(W)$ is the triangle containing v_1 fixed by hyperplane W in R_1, then $T(W) \subseteq S(U)$ if and only if $U \leq W$. In terms of distance-regularity we have $c_2 = 3$, $a_3 \geq 3$.*

Proof Let S be the orbit of v_1 under G_3. Then $|S| = [G_3 : G_{13}] = 15$ and G_3 acts on S with kernel T_3 inducing the action of $S_6 \cong G_3/T_3$ on 2-subsets of the associated 6-set with subdegrees 1, 6 and 8. Since $G_2 = G_{12}G_{123}$, the triangle T is contained in S, and since $[G_{13} : G_{123}] = 3$, three of the triangles at v_1 are contained in S. The subgroup R_{13} of order 4 is the kernel of R_1 acting on S. Hence the induced action is also of order $4 = |R_1/R_{13}|$. □

The subgroup G_{13} in G_1 is the stabilizer of the subspace $V_3 = \langle b_3, b_4, b_5 \rangle$ and hence the vertices in $S \cap (\{v_1\} \cup \Gamma_1(v_1))$ are identified with the non-zero vectors in V_3. To understand vertices at distance 3 from v_1 we start by establishing the following *quad neighbourhood* description.

Lemma 5.8 *The subgroup T_3 permutes transitively the 24 vertices in $\Gamma_1(v_1) \backslash S$ with a stabilizer that is elementary abelian of order 8.*

Proof The assertion of the lemma is an intrinsic property of $L_5(2)$, since in other words it claims that R_3 extended by an element $t \in K_3^+$ of order 3 which acts trivially on V_3 and permutes transitively the three hyperplanes containing V_3 is transitive on $V_5(2) \setminus V_3$. If a vector-vertex in $\Gamma_1(v_1) \setminus S$ is chosen to be b_2 then the corresponding stabilizer in R_3 is generated by the triple of transvections τ_{i1} taken for $i = 3$, 4 and 5. □

The *second quad neighbourhood* principle is the following.

Lemma 5.9 *Let $u \in S \setminus \Gamma_1(v_1)$ and $w \in \Gamma_1(v_1) \setminus S$. Then the stabilizer P_1 in G of the ordered triple (u, v_1, w) is isomorphic to $S_4 \times 2$.*

Proof By Lemma 5.7 S is the only quad containing u and v_1, thus S is stabilized by P_1 and the subgroup P_0 which stabilizes S vertex-wise along with w is a normal subgroup in P_1. On the other hand, $P_0 \cong 2^3$ by Lemma 5.8 and P_1/P_0 is the stabilizer of two non-adjacent vertices in a quad with respect to the natural action of the full automorphism group of the quad, which is S_6. Hence $P_1/P_0 \cong S_3$. It remains to show that the extension splits and that the action of the complement on P_0 is faithful. The calculations can be performed inside G_{13}, since P_1 stabilizes both v_1 and S, and as above we choose w to be b_2. The Sylow 3-subgroup of G_{13} is elementary abelian of order 9 generated by $t = \tau_{12}\tau_{21}$ and $s = \tau_{34}\tau_{43}$. Since t does not stabilise b_2, we conclude that P_1 contains s (which acts faithfully on P_0, since it does not centralize τ_{31}). If X is the subgroup in G_{13} which normalizes $\langle s\rangle$ and also stabilizes b_2 then

$$X = \langle s, \tau_{34}\rangle \times \langle \tau_{21}\rangle \cong S_3 \times 2,$$

so that the extension of P_0 by S_3 splits, hence $P_1 \cong S_4 \times 2$, proving the statement. □

Lemma 5.10 *Let $p = (v_1, x, y, z)$ be a path in Γ such that $y \in \Gamma_2(x)$ and z is not contained in the quad determined by v_1 and y. Let P be the stabilizer of z in G_1, let P_{12} be the stabilizer of p, and let P_1 and P_2 be the stabilizers in P of x and y, respectively. Then*

$$P_1 \cong P_2 \cong S_4 \times 2, \ \ [P_1 : P_{12}] = [P_2 : P_{12}] = 3 \text{ and } Z(P_1) \neq Z(P_2).$$

Proof With (u, v_1, w) as above, let t be one of the three common neighbours of u and v_1. Then t is contained in the quad S, while t and w are at distance 2 and hence are contained in a quad S', which differs from S since $w \notin S$. By the symmetric version of the *second quad neighbouring* description, if P_2 is the stabilizer in G of the ordered triple (u, t, w), then $P_2 \cong S_4 \times 2$ and, because of the triple choice for t, the intersection of P_1 and P_2 (denoted by P_{12}) has index 3 in each. Furthermore, we claim that P_1 and P_2 have different centres. In fact, from the structure of the subgroup X above we observe that $Z(P_1) \leq R_1$. By the symmetry $Z(P_2) \leq R_1(t)$, where $R_1(t)$ is the conjugate of R_1 under an element of G which maps v_1 onto t. Without loss of generality, we assume that $t \in T$, say $t = v_2$. Then both $Z(P_1)$ and $Z(P_2)$ are contained in G_2, while the structure of the latter shows that R_1 and $R_1(v_2)$ are disjoint and the result follows. □

By the symplectic identification Lemma 3.14 the information on P we have unearthed is sufficient to identify P inside G_1 up to isomorphism, which concludes the reconstruction of the octad graph from the Mathieu amalgam. Thus we have the following *reconstruction* principle.

Lemma 5.11 *With path p as above $z \in \Gamma_3(v_1)$, $P \cong Sp_4(2) \cong S_6$. There are 45 shortest paths joining v_1 with z, which are transitively permuted by P.* □

By the lemmas proved in this section, we conclude that Γ is determined uniquely up to isomorphism and that it is distance-transitive with the following intersection array:

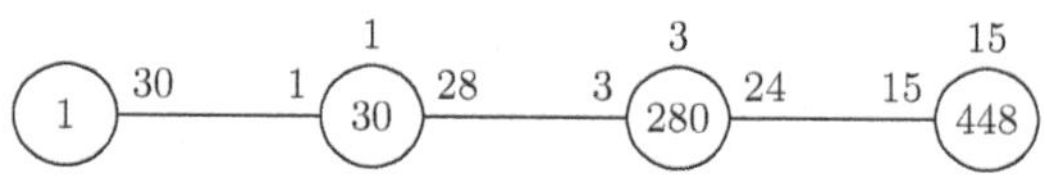

We summarize this conclusion in the following theorem.

Theorem 5.12 *Let $\mathcal{A} = \{G_1, G_2, G_3\}$ be the Mathieu amalgam, let G be a faithful completion of $\mathcal{A}$ and let Γ be the graph on the cosets of G_1 in G, where two vertices are adjacent whenever they intersect a common coset of G_2. Then*

(i) *G exists and is unique up to isomorphism;*
(ii) *the diameter of Γ is 3, and it is distance-transitive with the above diagram;*
(iii) *Γ has 759 vertices and it is uniquely determined by the amalgam $\mathcal{A}$;*
(iv) *G is the unique faithful completion of $\mathcal{A}$, which is thus the universal completion, and*

$$|G| = |G_1| \cdot 759 = 2^{10} \cdot 3^3 \cdot 5 \cdot 7 \cdot 11 \cdot 23.$$ □

The unique faithful completion G of the Mathieu amalgam is the Mathieu group M_{24} in its new incarnation, and some familiar properties of this group will be recovered in the subsequent sections of this chapter.

5.3 The Octad Graph on 8-Subsets

Every completion (of the Mathieu amalgam) can be factorized through a universal completion. Since the order of the faithful completion of the Mathieu amalgam has been proved to be uniquely determined, there is indeed only one faithful completion which is universal. In particular, the completion φ constructed in Section 5.1 is universal. The completion φ will be viewed as a realization of G as a permutation group on the 24-set $\mathcal{P}$. Since the representation is faithful by virtue of the construction, there are 759 images of $\mathcal{P}_8$ under G and the smallest graph on these 759 subsets of size 8 is precisely the octad graph.

Define a G-invariant mapping π of the vertex set of Γ (viewed as the coset graph) into the set of 8-subsets of $\mathcal{P}$ by

$$\pi(G_1 h) = \mathcal{P}_8^h,$$

where the right-hand side of the equality is the image of $\mathcal{P}_8$ under a representative h of the coset-vertex $G_1 h$. Clearly $\pi(v_1) = \pi(G_1) = \mathcal{P}_8$. The following *intersection* attribute is very important.

Lemma 5.13 *In the above terms if $u, v \in \Gamma$ and $u \in \Gamma_i(v)$ then*

$$|\pi(u) \cap \pi(v)| = 8,\ 0,\ 4, 2 \ \ for\ \ i = 0,\ 1,\ 2,\ 3,$$

respectively.

Proof If $i = 0$, 1 or 2 in the above statement, then u and v are contained in a common quad, and the assertion can be deduced by considering the sextet associated with the quad. For the case $i = 3$ we argue as follows. Let S be the quad stabilized by G_3 and let $Q_1, Q_2, \ldots, Q_6$ be the 4-subsets forming S, such that $Q_1 \cup Q_2 = \mathcal{P}_8$ and the remaining Q_is are the orbits on $\mathcal{P}_{16}$ of the subgroup R_{13}. Then $Q_1 \cup Q_3$ is $\pi(w)$ for a vertex w at distance 2 from v_1 contained in S. If $u \in \Gamma_1(v_1) \setminus S$ (so that $u \in \Gamma_3(w)$) then $\pi(u)$ is an orbit on $\mathcal{P}_{16}$ of a hyperplane H in R_1, such that R_{13} is not in H. Then $\pi(u) \cap \pi(w)$ is an orbit on $\mathcal{P}_{16}$ of the 1-subspace $H \cap R_{13}$, which is of size 2. □

If $z \in \Gamma_3(v_1)$ then we know that the stabilizer $G_1(z)$ of z in $G_1 \cong 2^4 : L_4(2)$ is isomorphic to S_6 and, since $|\pi(z) \cap \mathcal{P}_8| = 2$, it is easy to see that $G_1(z)$ acts on $\mathcal{P}$ with orbit lengths

$$2,\ 6,\ 6,\ 10.$$

On the other hand, an $L_4(2)$-complement to R_1 in G_1 acts on $\mathcal{P}$ with orbit lengths

$$1,\ 8,\ 15.$$

This gives the following *second complement* principle.

Lemma 5.14 *The stabilizer in G_1 of a vertex from $\Gamma_3(v_1)$ is an S_6-subgroup that is not contained in an $L_4(2)$-complement.* □

Notice that, while the class of $L_4(2)$-subgroups in G_1 is unique by Lemma 4.5, there are two classes of S_6-subgroups, since there exists a unique indecomposable extension of R_1 (viewed as an S_6-module) by the trivial one-dimensional module. This extension appears in the following way. Consider one of the two degree-6 permutation representations of S_6 such that R_1 is the heart of the corresponding $GF(2)$-permuation module. Then the required extension is the permutation module factorized over the trivial one-dimensional submodule.

We conclude this section with the following *vertex residue* structure which will play an important role in the subsequent exposition, and is an immediate consequence of the structure of the vertex stabilizer $G_1 \cong 2^4 : L_4(2)$ and its intersections with the stabilizer G_2 of the triangle T and the quad stabilizer G_3.

Lemma 5.15 *The rank-2 geometry of triangles and quads containing a given vertex (say the one stabilized by G_1) of the octad graph is isomorphic to the geometry of one- and two-dimensional subspaces of the natural module of $G_1/R_1 \cong L_4(2)$ dual to R_1, and this isomorphism can be taken to commute with the natural actions of G_1.* □

5.4 Simplicity, 5-Transitivity and the Steiner Property

We start by proving the Steiner property (see Section 1.3) for the pair $(\mathcal{P}, \mathcal{B})$, where $\mathcal{P}$ comes from the completion φ of the Mathieu amalgam of degree 24 and $\mathcal{B}$ is the set on images of $\mathcal{P}_8$ under G (viewed as the group generated by the image of φ). Alternatively $\mathcal{B}$ can be defined as the image of the vertex set of Γ under the projection map π defined in the previous section. The elements of $\mathcal{B}$ can rightfully be called *octads*. The group G permutes the octads transitively, with $G_1 \cong 2^4 : L_4(4)$ being the stabilizer of the octad $\mathcal{P}_8$, where G_1 acts on $\mathcal{P}_8$ as the alternating group of this set, while on $\mathcal{P}_{16}$ the subgroup G_1 acts as on the cosets of an $L_4(2)$-complement. Because of the intersection property, two distinct octads never share a 5-subset of $\mathcal{P}$; on the other hand, each of the 759 octads contains $\binom{8}{5}$ subsets of size 5. Directly from the *miracle equality*

$$\binom{8}{5} \cdot 759 = \binom{24}{5}$$

we deduce the following *Steiner property* of $(\mathcal{P}, \mathcal{B})$.

Lemma 5.16 *Every 5-subset of $\mathcal{P}$ is contained in a unique octad, so that $(\mathcal{P}, \mathcal{B})$ is a Steiner system of type $S(24, 8, 5)$.* □

Towards the proof of 5-fold transitivity of G on $\mathcal{P}$, consider two 5-subsets Q_1 and Q_2 of $\mathcal{P}$ equipped with some orderings α_1 and α_2, respectively. Let $B(Q_1)$

and $B(Q_2)$ be the uniquely determined octads containing these subsets. Then, by the transitivity of G on octads, there are $g_1, g_2 \in G$ such that

$$(Q_1^{g_1}, \alpha_1^{g_1}) \text{ and } (Q_2^{g_2}, \alpha_2^{g_2})$$

are two ordered 5-subsets of $\mathcal{P}_8$. Since G_1 induces on $\mathcal{P}_8$ the alternating group of this set, there is an element in G_1 which maps one of the ordered 5-sets onto another one, and we have proved the following.

Lemma 5.17 *The permutation group* $(G, \mathcal{P})$ *is 5-fold transitive.* □

It is worth mentioning and not too hard to prove that G acts on the set of 6-subsets in $\mathcal{P}$ with two orbits. A representative of one orbit can be found inside the octad $\mathcal{P}_8$, and its stabilizer is a subgroup of G_1 of the form $2^4 : S_6$; a representative of the other orbit intersects $\mathcal{P}_8$ in a 5-set, its stabilizer has the form $3 \cdot S_6$, and it is conjugate to a complement to R_3 in G_3.

Although the simplicity of G can be proved in one line, we take the opportunity to discuss the simplicity issue in the amalgam setting. Let $\mathcal{X} = \{A_i \mid 1 \leq i \leq n\}$ be an amalgam of rank n, let A be a faithful completion of $\mathcal{X}$ and let N be a normal subgroup of A. There are certain constraints on how N might intersect $\mathcal{X}$. The constraints come from the following two observations: (1) $N_i := N \cap A_i$ is a normal subgroup of A_i; and (2) if K_i is a normal subgroup of A_i contained in N_i then the normal closure in A_j of $K_i \cap A_j$ is contained in N_i, where $1 \leq i, j \leq n$. Let us say that a normal subgroup K_i in A_i *implies* a normal subgroup K_j in A_j if K_j is contained in the normal closure of $K_i \cap A_j$ in A_j. From the above observations, whenever a normal subgroup N of A contains K_i, it contains every subgroup K_j which is implied by K_i. We say that the amalgam $\mathcal{X}$ as above is *essentially simple* if for every $1 \leq i, j \leq n$ and every normal subgroup K_i of A_i there is a sequence

$$(K_{m(1)} = K_i, K_{m(2)}, \ldots, K_{m(l)} = A_j)$$

such that $K_{m(k)}$ implies $K_{m(k+1)}$ for every $1 \leq k \leq l - 1$. It is immediately evident that for an essentially simple amalgam every proper normal subgroup in a faithful completion of the amalgam intersects the amalgam trivially. It is an elementary exercise to check that the Mathieu amalgam is essentially simple.

Now suppose that $\mathcal{X}$ is essentially simple, that A is a finite faithful completion of $\mathcal{X}$, and that N is a proper normal subgroup in A (as usual we identify $\mathcal{X}$ with its image in A). Then $N \cap \mathcal{X} = \{1\}$, which particularly implies that N acts semi-regularly on the vertex set of the coset graph of A, which provides us with the following congruences on $|N|$:

$$|N| \text{ divides } [A : A_i] \text{ for every } 1 \leq i \leq n.$$

Furthermore, unless $|N| = [A\colon A_i]$, the orbits of N on the set of cosets of A_i in A are the classes of an equivalence relation on A/A_i preserved by A.

The primitivity of the action of G on the vertex set of the octad graph Γ is immediately evident from the intersection diagram of Γ. Thus a normal subgroup of G can only be of order 759 (which already is nonsensical), and this number does not divide

$$|\mathcal{S}| = [G : G_3] = \frac{759 \cdot 35}{15},$$

which gives us the final contradiction regarding the existence of N. Alternatively, considering the permutation representation φ of degree 24 of G, we observe that the representation is primitive, since it is 5-fold transitive. Therefore a non-identity normal subgroup N of G must act transitively on $\mathcal{P}$ and hence intersect properly (every) Sylow 2-subgroup in G. Since such a subgroup can be found inside G_{123}, owing to the essential simplicity of $\mathcal{A}$ the normal subgroup N must be the whole of G. Thus we have established the simplicity of the group M_{24} in its new incarnation.

Theorem 5.18 *The universal completion group G of the Mathieu amalgam is simple.* □

5.5 Octad Space and the Golay Code

Now when $(\mathcal{P}, \mathcal{B})$ has been proved to be a G-invariant Steiner system of type $S(24, 8, 5)$ an easy construction of the Golay code is immediately available: just take $2^{\mathcal{P}}$ to be the $GF(2)$-vector space with basis $\mathcal{P}$ and define $\mathcal{C}$ to be the linear span of $\mathcal{B}$ in $2^{\mathcal{P}}$. The conditions (a) to (d) in Section 1.1 follow from Lemma 5.13, which ensures that any two octads intersect evenly. Towards (e) and (f) we calculate the *weight enumerator* of $\mathcal{C}$, which is a rule which gives the number m_l of codewords of any given length l.

The subspace $\mathcal{C}$ is totally singular with respect to the non-degenerate symplectic form

$$(A, B) \mapsto |A \cap B| \mod 2$$

on $\mathcal{P}$ and hence

$$\sum_{l=0}^{24} m_l = |\mathcal{C}| \leq 2^{\frac{|\mathcal{P}|}{2}} = 2^{12}.$$

On the other hand, we know that $m_0 = 1$ (as always), $m_{24} = 1$ (since trios exist), $m_8 \geq 759$ (for the octads) and $m_{16} \geq 759$ (for the octad complements). Thus it remains only to prove the following *dodecad* property.

Lemma 5.19 *Let a* dodecad *be defined as the sum of two octads intersecting in a pair. Then exactly* $66 = \binom{12}{2}$ *such sums give the same dodecad and the complement of a dodecad is again a dodecad, so that if* $\mathcal{D}$ *is the set of dodecads then*

$$|\mathcal{D}| = \frac{759 \cdot 448}{132} = 2576. \qquad \square$$

Proof According to the paragraph preceeding the lemma, $|\mathcal{D}| \leq m_{12} \leq 2576$, thus all we need to show is that, given four octads B_1, B_2, B_3, B_4 with $B_1 \cap B_2$ and $B_3 \cap B_4$ being the same subset of size 2, the equality

$$B_1 + B_2 = B_3 + B_4$$

holds. Suppose the contrary. Since $B_1 \subseteq B_3 \cup B_4$, while $B_1 \cap B_3$ and $B_1 \cap B_4$ are two even sets, both containing $B_3 \cap B_4$, it is unavoidable that one of the octad intersections will be of size at least 6, which is not possible. $\square$

By Lemma 5.19 and the calculations preceding it, the weight enumerator of the Golay code $\mathcal{C}$ is the following:

$$m_0 = 1, \ m_8 = 759, \ m_{12} = 2576, \ m_{16} = 759, \ m_{24} = 1.$$

5.6 Maximal Parabolic Geometry

The members of the Mathieu amalgam $\mathcal{A}$ are maximal 2-local subgroups in $G \cong M_{24}$ (in the sense that they are maximal subject to the condition of being the normalizers of 2-subgroups). The coset graph of G with respect to $\mathcal{A}$ has vertices identified with octads, trios and sextets, and it is the incidence graph of the Ronan–Smith maximal 2-local parabolic geometry $\mathcal{G}(M_{24})$ of M_{24}.[1] The geometry $\mathcal{G}(M_{24})$ is a so-called locally truncated C_4-geometry.

A typical example of C_4-geometry is the one associated with the group $Sp_8(2)$, which is the automorphism group of an eight-dimensional $GF(2)$-space $V_8(2)$ equipped with a non-singular symplectic form f. The elements of the C_4-geometry $\mathcal{G}(Sp_8(2))$ are the subspaces in $V_8(2)$ which are totally singular with respect to f; the type $t(U)$ of such a subspace U is determined by its dimension, and for our present purpose it is convenient to set

$$t(U) = 5 - \dim(U).$$

[1] M. A. Ronan and S. D. Smith, 2-Local geometries for some sporadic groups, in *Proceedings of Symposia in Pure Mathematics 37 (Finite Groups)*, American Mathematical Society, 1980, pp. 283–289.

The incidence is the symmetrized inclusion relation. Then $\mathcal{G}(Sp_8(2))$ belongs to the C_4-diagram

The nodes correspond to types indicated above the nodes. A *flag* is a set of pairwise incident subspaces. A maximal flag contains one element of each type, and every flag Ψ is contained in (possibly more than) one maximal flag. The set of types of elements needed to complete Ψ to a maximal flag is said to be the *co-type* of Ψ. The number q_i placed under the node of type i shows that every pre-maximal flag of co-type i is contained in $q_i + 1$ maximal flags. The edge between the nodes of type i and j indicates the rank-2 geometry of the pre-maximal flags containing a flag of co-type $\{i, j\}$ (which is known as the residue of that flag). According to the standard notation we place the empty edge for the complete geometry, where any two elements of different types are incident, a single edge for the projective plane, and a double edge for the geometry of one- and two-dimensional totally singular subspaces in a four-dimensional symplectic space (a *quad*).

The geometry $\mathcal{G}(Sp_8(2))$ is a dual polar space, where the term 'dual' reflects the convention that larger subspaces have smaller types. It is instructive to take a look at the collinearity graph Σ of that geometry to notice its analogy with the octad graph. By the definition the vertices of Σ are the elements of type 1 in the geometry (they correspond to the maximal totally singular subspaces in $(V_8(2), f)$, which are four-dimensional). Two vertex-subspaces are adjacent whenever their intersection has co-dimension 1 in each, i.e. whenever they are incident to a common element of type 2 in the geometry. The graph Σ is distance-regular with the following intersection diagram (which bears a striking similarity to the diagram of the octad graph):

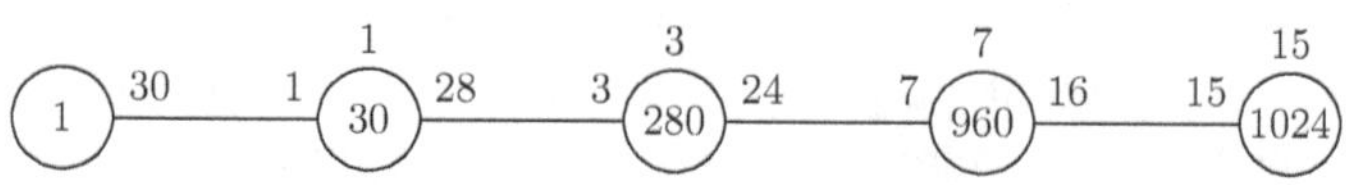

In order to demonstrate a deeper analogy we can redefine $\mathcal{G}(Sp_8(2))$ in terms of so-called geometric subgraphs in Σ. A geometric subgraph of type i (where $1 \leq i \leq 4$) is the subgraph induced by the vertex-subspaces incident to a given element of type i in the geometry). The subgraphs of types 1, 2 and 3 are, respectively, the vertices, the triangles and the quads in Σ, while those of type 4 are the distance-regular subgraphs with the following intersection diagram:

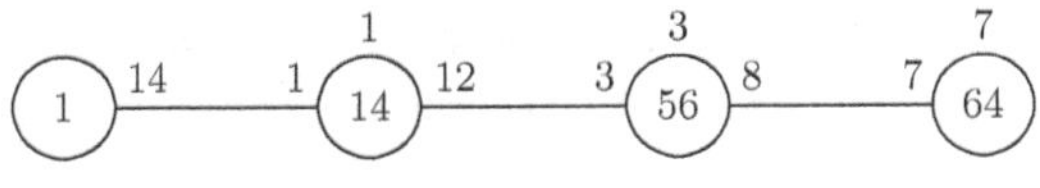

A pair of vertices at distance j in Σ (where $0 \le j \le 3$) is contained in a unique geometric subgraph of type $j+1$. Furthermore, for every vertex x of Σ the geometric subgraphs of type greater than 1 containing x form the projective geometry of the four-dimensional subspace-vertex x. In terms of the geometric subgraphs, the incidence relation in $\mathcal{G}(Sp_8(2))$ is still via the symmetric inclusion.

Now, if we were to *truncate* the geometry $\mathcal{G}(Sp_8(2))$ by removing the elements of type 4, we would obtain a rank-3 geometry with the diagram

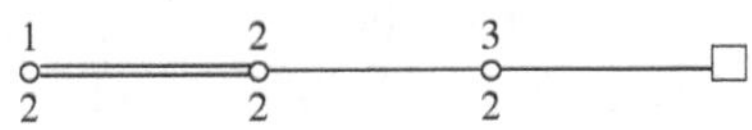

where the leftmost box stands for a fake node, which saves us from inventing a new type of edge for the rank-2 geometry of one- and two-dimensional subspaces in a four-dimensional $GF(2)$-space.

The crucial observation, which follows from the *vertex residue* structure presented in Lemma 5.15, is that the geometry $\mathcal{G}(M_{24})$ of vertices, triangles and quads in the octad graph is also described by the above *locally truncated* C_4-*geometry* diagram. As was pointed out in the introduction, the characterization by M. Ronan[2] of $\mathcal{G}(M_{24})$ in the class of locally truncated C_4-geometries was the main motivation for our construction of M_{24}. In fact, by Ronan's result Theorem 1.10 the geometry $\mathcal{G}(M_{24})$ is simply connected and thus M_{24} is the universal completion of the amalgam of its maximal 2-local subgroups. Within the current exposition the universality is proved in Theorem 5.12.

The prototype for our construction of the Mathieu amalgam was the group $H = GL_5(2)$ with its rank-3 amalgam $\mathcal{H} = \{H_1, H_2, H_3\}$ formed by the stabilizers in H of one-, two- and three-dimensional subspaces from the natural module $V_5(2)$ of H. The coset geometry of $\mathcal{H}$ in H has diagram

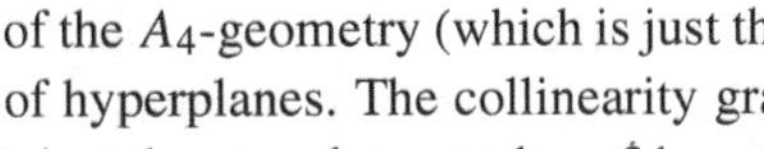

and it is a truncation of the A_4-geometry (which is just the projective geometry of $V_5(2)$) by the set of hyperplanes. The collinearity graph is not particularly spectacular, since it is just the complete graph on 31 vertices and the geometry cannot be recovered from this graph in purely combinatorial terms. In fact, the automorphism group of the collinearity graph, being the symmetric group

[2] M. A. Ronan, Locally truncated buildings and M_{24}, *Math. Z.* **180** (1982), 469–501.

of degree 31, is larger than the automorphism group $L_5(2)$ of the projective geometry. At this stage it is natural to ask the following question.

Question 5.20 *Is there a C_4-geometry of which $\mathcal{G}(M_{24})$ is the truncation with respect to elements of type 4?*

The negative answer to Question 5.20 justified in the next section is in a sense very 'positive', since it leads to the *minimal 2-local parabolic geometry*[3] $\widetilde{\mathcal{G}}(M_{24})$ of M_{24}. The diagram

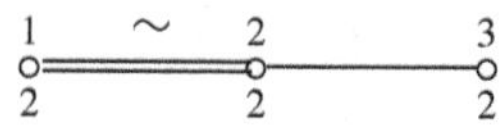

of $\widetilde{\mathcal{G}}(M_{24})$ does have fake nodes, but it involves a semi-classical *tilde* geometry that is the triple cover of the S_6-quad, which is related to the non-split extension $3 \cdot S_6$. Notice that the tilde geometry is the geometry of vertices in two bipartite parts of the famous *Foster graph*.

5.7 Tilde Geometry of M_{24}

Let $\mathcal{G}(M_{24})$ be the above-constructed locally truncated C_4-geometry, which is the coset geometry of the Mathieu amalgam $\mathcal{G}$ completed in M_{24}.

Lemma 5.21 *The octad graph* Γ does not *contain a family $\mathcal{X}$ of valency* 14-*subgraphs such that for every vertex x of Γ and for every hyperplane h in the geometry associated with x in the sense of Lemma* 5.15 *there is a unique subgraph $\Xi = \Xi^{(x,h)}$ in $\mathcal{X}$ which satisfies the equality $\Xi(x) = h$.*

Proof Let x be the vertex of Γ stabilized by G_1. Then the residual geometry of x can be described in terms of subsets of the set $\Gamma(x)$ of neighbours of x in Γ. The points are triangles on x with x removed, while a line is the set of six neighbours of x in a quad on x. For $r \in R_1^{\#}$ the set $h(r)$ of 14 vertices in $\Gamma(x)$ fixed by r is a hyperplane, and all the hyperplanes can be obtained in this way. Since G_1 acts transitively on $R_1^{\#}$ and G acts vertex-transitively on Γ, the family $\mathcal{X}$ must be G-invariant. Let S be the quad stabilized by G_3. Then the set $S(x)$ of neighbours of x in S is the fixed points of R_{13} in $\Gamma(x)$. Since R_{13} is of order 4, the set $S(x)$ is contained in precisely three hyperplanes (which are $h_1 = h(\tau_{51})$, $h_2 = h(\tau_{52})$ and $h_{1\frac{1}{2}} = h(\tau_{51}\tau_{52})$), transitively permuted by G_{13}. Therefore, S is contained in just three members of $\mathcal{X}$, which are $X^{(u,h_j)}$

[3] M. A. Ronan and G. Stroth, Minimal parabolic geometries for the sporadic groups, *Europ. J. Combin.* **5** (1984), 59–91.

for $j = 1$, 2 and $1\frac{1}{2}$. These three members must be transitively permuted by G_3, which is impossible, since

$$G_3 \cong 2^6 : 3 \cdot S_6$$

does not contain subgroups of index 3 (since $G_3/R_3 \cong 3 \cdot S_6$ is a non-split extension). Thus $\mathcal{G}(M_{24})$ is not a truncation of a C_4-geometry and the answer to Question 5.20 is negative. □

Let us now modify the argument to get a positive outcome. Let G_{14} be the stabilizer of $h(\tau_{51})$ in G_1, then G_{14} is the stabilizer of a point–hyperplane flag in the action of $H \cong L_5(2)$ on the projective geometry of $V_5(2)$ and

$$G_{14} = C_{G_1}(\tau_{51}) \cong 2_+^{1+6} : L_3(2).$$

Let T be the triangle in the quad graph stabilized by G_2, and let $S^{(1)}$, $S^{(2)}$ and $S^{(3)}$ be the quads containing T. Then

$$h(x) = \bigcup_{i=1}^{3} S^{(i)}(x)$$

is a hyperplane in the residue of $x \in T$, and the subgraph $\Xi^{(x,h(x))}$, as in Lemma 5.21, must be independent of the choice of x as long as x is in T. Thus we define

$$G_{24} = N_{G_2}(G_{14} \cap G'_{12}) \cong 2^6.(S_4 \times S_3).$$

Then, if $\mathcal{X}$ existed, the subamalgam $\mathcal{A}_4 := \{G_{14}, G_{24}\}$ would be formed by stabilizers of x and T in the stabilizer of $\Xi^{(x,h(x))}$. Furthermore, the latter subgraph (and therefore the whole family $\mathcal{X}$) exists if and only if $\mathcal{A}_4$ generates in G a proper subgroup. We already know that this is not the case, and the justification can be restated as follows. By intersecting $\mathcal{A}_4$ with G_3 we obtain a subamalgam $\mathcal{A}_{34}$ whose core is R_3, and the quotient amalgam is an $\{S_4 \times 2, S_4 \times 2\}$-subamalgam in $G_3/R_3 \cong 3 \cdot S_6$ (which is the amalgam G_3^1 from Goldschmidt's list). This quotient amalgam generates the whole of $G_3/R_3 \cong 3 \cdot S_6$ because of the non-splitness, and therefore $\mathcal{A}_4$ generates the whole of G. Thus we have the following.

Theorem 5.22 *The group $G \cong M_{24}$ is a generated completion of the subamalgam*

$$\widetilde{\mathcal{A}} = \{G_{14}, G_{24}, G_3\}$$

of the Mathieu amalgam and the corresponding coset geometry is the tilde geometry $\widetilde{\mathcal{G}}(M_{24})$ *with the diagram*

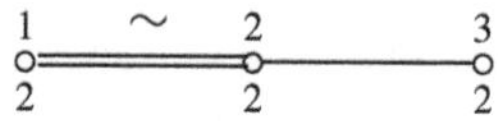

□

Notice that the coset geometry of the subamalgam $\{H_{14}, H_{24}, H_3\}$ in the $L_5(2)$-amalgam is not *residually connected*,[4] since

$$\langle H_{134}, H_{234}\rangle = H_{34} \neq H_3.$$

[4] A. Pasini, *Diagram Geometries*, Oxford University Press, Oxford, 2006.

6

Maximal Subgroups

A maximal subgroup X of M_{24} usually preserves a nice additional sub-structure on the Golay code $\mathcal{C}$, on the Steiner system $(\mathcal{P}, \mathcal{B})$ and/or on the octad graph Γ. By virtue of the maximality of X it is the full stabilizer in M_{24} of the relevant sub-structure. Often one can construct the sub-structure by making use of some natural intrinsic terms of X. Commonly the global object appears as a union of some X-orbits which have to be glued together in a sophisticated way in order to satisfy the required axioms. To get the whole of M_{24} one has to produce an automorphism of the constructed structure which does not preserve the sub-structure and thus lies outside X (this is usually the most difficult step in the construction). The existence of such an automorphism ensures the smoothness of the gluing. Most of the known constructions of M_{24} and of the other Mathieu groups fall within this framework. A construction from a maximal subgroup demonstrates the existence of the relevant sub-structure and identifies its stabilizer. Some of these constructions are also useful for uniqueness proofs for the global structures. Overall the Mathieu group M_{24} has nine classes of maximal subgroups.[1,2] If we arrange the representatives of these classes in decreasing order and denote by $G^{(i)}$, $1 \leq i \leq 9$, the ith representative in that order then the list of the isomorphism types of the maximal subgroups is as follows:

$$M_{23},\ M_{22}:2,\ 2^4:L_4(2),\ M_{12}:2,\ 2^6:3\cdot S_6,$$
$$L_3(4):S_3,\ 2^6:(L_3(2)\times S_3),\ L_2(23),\ L_2(7).$$

In particular, $G_1 = G^{(3)}$, $G_2 = G^{(7)}$ and $G_3 = G^{(5)}$. In this chapter we devote a separate section to every maximal subgroup which is not a member

[1] C. Choi, On subgroups of M_{24}. I: Stabilizers of subsets, *Trans. Amer. Math. Soc.* **167** (1972), 1–27.

[2] C. Choi, On subgroups of M_{24}. II: The maximal subgroups of M_{24}. *Trans. Amer. Math. Soc.* **167** (1972), 29–47.

of the Mathieu amalgam (with the exception of $G_2 = G^{(7)}$, which is implicit in Turyn's construction presented in Section 6.5).

6.1 The Point Stabilizer M_{23}

In this section we discuss the stabilizer $G^{(1)}$ in $G \cong M_{24}$ of a point $p \in \mathcal{P}$. This stabilizer is another Mathieu group M_{23} of order

$$|M_{23}| = 2^{10} \cdot 3^3 \cdot 5 \cdot 7 \cdot 11,$$

where the index indicates the size of the complement to p in $\mathcal{P}$ on which $G^{(1)}$ acts 4-fold transitively. Since G acts primitively on $\mathcal{P}$, the subgroup $G^{(1)} \cong M_{23}$ is manifestly maximal in G with index 24. In what follows we describe the action of $G^{(1)}$ on the octad graph and realize $G^{(1)}$ as a completion of a certain subamalgam in $\mathcal{A}$. This completion happens to be universal.

Since G acts transitively on $\mathcal{P}$, the particular choice of p is irrelevant and we take $p \in \mathcal{P}_{16}$. Then

$$G_1^{(1)} := G^{(1)} \cap G_1$$

is an $L_4(2)$-complement to R_1 in G_1 and without loss of generality we take it to be the Levi complement K_1. Therefore $G_1^{(1)}$ fixes $u = \pi^{-1}(\mathcal{P}_8)$ and acts on the set of triangles containing u as on the set of non-zero vectors of the dual natural module R_1^* of $G_1^{(1)} \cong L_4(2)$. In this case the intersection

$$G_{12}^{(1)} := G^{(1)} \cap G_{12}$$

is the maximal parabolic in K_1, equal to

$$R_2^{(4)} L = R_2^{(4)} L^{(4)} \cong 2^3 : L_3(2).$$

According to Table 5.1 $p \in \mathcal{P}_8^{(4)}$. Furthermore, the normalizer $G_2^{(1)}$ of $G_{12}^{(1)}$ in G_2 is an extension of $G_{12}^{(1)}$ by the restriction of an automorphism β of G'_{12} which performs the permutation

$$(L\, L^{(4)})(L^{(5)})(L^{(4\frac{1}{2})})$$

on the classes of $L_3(2)$-complements (it should be remembered that G_2 normalizes the class of $L^{(5)}$). The automorphism β cannot be realized by an element of G_1, thus the corresponding element of $G_2^{(1)}$ maps u onto another vertex within the triangle T stabilized by G_2, although $G_2^{(1)}$ still stabilizes the point $p \in \mathcal{P}$. Thus we have found in $\{G_1, G_2\}$ a subamalgam

$$\{G_1^{(1)}, G_2^{(1)}\} \cong \{L_4(2), 2^3 : L_3(2).2\} \cong \{L_4(2), \mathrm{Aut}(2^3 : L_3(2))\}$$

and $G^{(1)}$ can be redefined as the subgroup in G generated by this subamalgam. Since p is stabilized by the subamalgam, the equality $G^{(1)} \cap G_1 = G_1^{(1)} \cong L_4(2)$ holds also in the new definition.

We follow the usual procedure and define $G_3^{(1)}$ to be the subgroup in G_3 generated by the intersection of the latter with the subamalgam $\{G_1^{(1)}, G_2^{(1)}\}$. To identify the structure of $G_3^{(1)}$, notice first that $R_3 \cap G^{(1)} = R_3 \cap G_1^{(1)}$ has order 2^4 and is generated by the transvections τ_{31}, τ_{32}, τ_{41} and τ_{42}. The whole of $G_3^{(1)}$ is the normalizer of $R_3 \cap G^{(1)}$ in G_3, so that

$$G_3^{(1)} \cong 2^4 : (S_3 \times S_5)^+.$$

The subgraph $\Gamma^{(1)}$ of the octad graph induced by the octads disjoint from the fixed point p is M_{23}-distance transitive with the following intersection array:

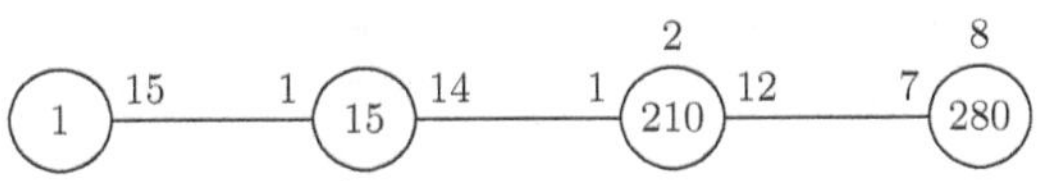

The subgroups $G_1^{(1)}$, $G_2^{(1)}$ and $G_3^{(1)}$ are the stabilizers in $G^{(1)} \cong M_{23}$ of a vertex, an edge and a Petersen subgraph in $\Gamma^{(1)}$, respectively.
The following result has been proved by S. V. Shpectorov and the author.[3]

Theorem 6.1 *The following three equivalent assertions hold:*

(i) *the fundamental group of the graph $\Gamma^{(1)}$ is generated by the homotopy classes of 5-cycles;*
(ii) *the coset geometry of $G^{(1)}$ with respect to the amalgam*

$$\mathcal{A}^{(1)} = \{G_1^{(1)}, G_2^{(1)}, G_3^{(1)}\} \cong \{L_4(2),\ 2^3 : L_3(2) : 2,\ 2^4 : (S_3 \times S_5)^+\}$$

is simply connected;
(iii) *$G^{(1)} \cong M_{23}$ is the universal completion group of the amalgam $\mathcal{A}^{(1)}$.* □

6.2 The Pair Stabilizer $M_{22}.2$

If $\{p, q\}$ is a two-element subset of $\mathcal{P}$ then its element-wise stabilizer in M_{24} is another sporadic simple Mathieu group M_{22} of order

$$|M_{22}| = 2^7 \cdot 3^2 \cdot 5 \cdot 7 \cdot 11,$$

[3] A. A. Ivanov and S. V. Shpectorov, The P-geometry for M_{23} has no nontrivial 2-coverings, *Europ. J. Combin.* **10** (1990), 347–362.

while the set-wise stabilizer $G^{(2)}$ is a maximal subgroup of M_{24} isomorphic to an extension of M_{22} by an automorphism of order 2 ($G^{(2)}$ is the automorphism group of M_{22}).

We choose $\{p, q\}$ to be contained in one of the 4-subsets of the sextet stabilized by G_3 outside of $\mathcal{P}_8$, put $G_i^{(2)} = G^{(2)} \cap G_i$ for $= 1, 2$ and 3, and form the subamalgam

$$\mathcal{A}^{(2)} = \{G_1^{(2)}, G_2^{(2)}, G_3^{(2)}\}.$$

Define $\Gamma^{(2)}$ to be the subgraph of the octad graph Γ induced by the images of $\mathcal{P}_8$ under $G^{(2)}$. By construction the vertex set of $\Gamma^{(2)}$ is formed by the octads disjoint from the pair $\{p, q\}$ with two vertex-octads adjacent whenever they are disjoint. The following result served as the starting point of the theory of Petersen and tilde geometries developed with S. V. Shpectorov, covered by the previous books cited in the Preface, and can be deduced from some basic properties of the Mathieu amalgam and of the octad graph.

Lemma 6.2 *The following assertions hold:*

(i) $\mathcal{A}^{(2)} \cong \{2^3 : L_3(2) \times 2,\ 2^4 : (2 \times S_4),\ 2^5 : S_5\}$;
(ii) *the subgroup* $G^{(2)} \cong M_{22}.2$ *acts distance-transitively on the subgraph* $\Gamma^{(2)}$;
(iii) *the intersection diagram of* $\Gamma^{(2)}$ *is the one given below the lemma;*
(iv) *the coset geometry associated with the embedding of* $\mathcal{A}^{(2)}$ *in* $G^{(2)}$ *is canonically isomorphic to the geometry of vertices, edges and Petersen subgraphs in* $\Gamma^{(2)}$. □

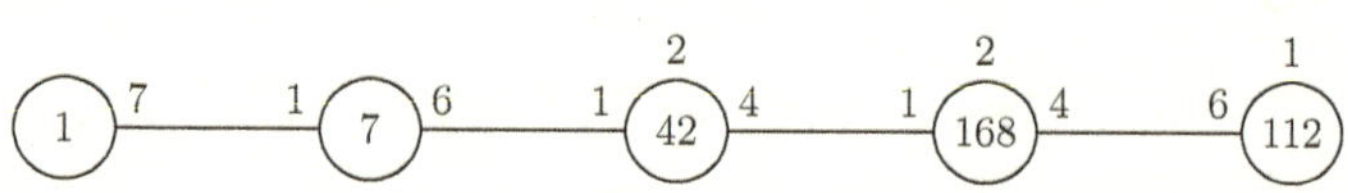

The geometry $\mathcal{G}(M_{22})$ of $G^{(2)}$ with respect to $\mathcal{A}^{(2)}$ is the rank-3 Petersen geometry with the diagram

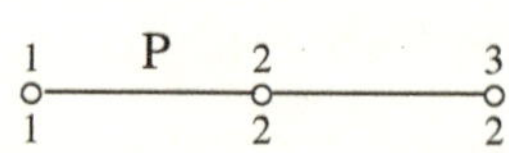

where the leftmost edge stays for the geometry of vertices and edges of the Petersen graph with the natural incidence relation. The coset geometry $\mathcal{G}(M_{23})$ of $G^{(1)}$ with respect to the amalgam $\mathcal{A}^{(1)} \cup (G^{(1)} \cap G^{(2)})$ (under the assumption that the point stabilized by $G^{(1)}$ is contained in the pair stabilized by $G^{(2)}$) is a rank-4 P-geometry with the diagram

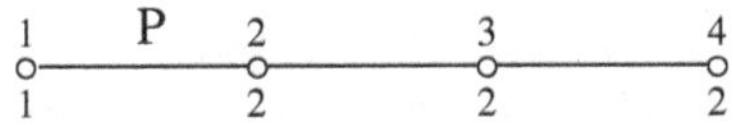

and it contains $\mathcal{G}(M_{22})$ as the residue of an element of type 4.

If x is a vertex-octad of the octad graph Γ and $x \in \Gamma^{(2)}$ then the trio-triangle of Γ containing along with x a vertex from $\Gamma_1^{(2)}(x)$ is a hyperplane in the residual geometry associated with x in the sense of Lemma 5.15. Therefore, we can make the following important observation.

Lemma 6.3 *The geometry $\mathcal{G}(M_{22})$ is a subgeometry in the tilde geometry $\widetilde{\mathcal{G}}(M_{24})$ and $\mathcal{A}^{(2)}$ is a subamalgam in the amalgam $\widetilde{\mathcal{A}} = \{G_{14}, G_{24}, G_3\}$.* □

Theorem 6.4 *The following assertions hold:*

(i) *in the fundamental group of the graph $\Gamma^{(2)}$ the homotopy classes of the 5-cycles generate a subgroup of index 3;*
(ii) *the covering graph of $\Gamma^{(2)}$ corresponding to the subgroup in the fundamental group generated by the homotopy classes of 5-cycles is the distance-transitive aptipodal triple cover of $\Gamma^{(2)}$ known as the Faradjev–Ivanov–Ivanov graph;*
(iii) *the universal completion of the amalgam $\mathcal{A}^{(2)}$ is isomorphic to $3 \cdot M_{22}.2$.* □

The commutator subgroup of the group $3 \cdot M_{22}.2$ in the above lemma is a non-split extension of M_{22} by a centre of order 3; the quotient over that centre of the whole group is $M_{22}.2$ and the elements outside the commutator subgroup invert the order-3 centre.

6.3 The Triple Stabilizer $P\Gamma L_3(4)$

Let X be a 3-subset of the 24-set $\mathcal{P}$, and let $G^{(3)}$ be the set-wise stabilizer of X in G. Because of the 5-fold transitivity of G on $\mathcal{P}$, the choice of X does not effect $G^{(3)}$ up to conjugacy. Furthermore, the structure of $G^{(3)}$ and the way it acts on $(\mathcal{P}, \mathcal{B})$ can be deduced directly from the axioms of the Steiner system and can be rearranged into a uniqueness proof for the system.

Set

$$\mathcal{Q} = \mathcal{P} \setminus X, \quad \mathcal{E} = \{E \mid E = B \setminus X \text{ for } B \in \mathcal{B}\},$$

and define

$$\varphi : \mathcal{E} \to 2^X$$

by the condition that $E \cup \varphi(E) \in \mathcal{B}$ for every $E \in \mathcal{E}$. Since a 5-subset is contained in a unique octad from $\mathcal{B}$ the subsets in $\mathcal{E}$ are pairwise different and φ is well defined. Further put

$$\mathcal{E}^Y = \{E \mid E \in \mathcal{E},\ \varphi(E) = Y\},\quad \mathcal{E}^{(k)} = \{E \mid E \in \mathcal{E},\ |\varphi(E)| = k\}.$$

Then $\mathcal{E}^{(3)}$ is easily seen to contain exactly 21 subsets of $\mathcal{Q}$ with size 5 each; any two elements from $\mathcal{Q}$ are in a unique subset from $\mathcal{E}^{(3)}$; if $E_1, E_2 \in \mathcal{E}^{(3)}$ then $E_1 \cup \varphi(E_1)$ and $E_2 \cup \varphi(E_2)$ are octads which share X, hence they intersect in a 4-set, which implies that $|E_1 \cap E_2| = 1$, so that a given element of $\mathcal{Q}$ is in 5-subsets from $\mathcal{E}^{(3)}$, while $\mathcal{E}^{(3)}$ is formed by 5-subsets of $\mathcal{Q}$, where the latter is a 21-set. This demonstrates the following.

Lemma 6.5 *The pair* $\Xi = (\mathcal{Q}, \mathcal{E}^{(3)})$ *is a projective plane of order* 4. □

As pointed out in Section 3.11, there is a unique projective plane of order 4 and therefore Ξ is the geometry of one- and two-dimensional subspaces in a three-dimensional $GF(4)$-space $V_3(4)$ and we can explore the properties of the plane illuminated in that section. It is also clear that $G^{(3)}$ acts faithfully on $\mathcal{E}$ and hence

$$\frac{|G|}{\binom{24}{3}} = |G^{(3)}| \leq |\mathrm{Aut}(\Xi)|.$$

Since the equality is attained, we conclude that

$$G^{(3)} \cong \mathrm{Aut}(\Xi) \cong P\Gamma L_3(4).$$

The subgroup $G^{(3)}$ possesses a surjective homomorphism onto

$$\mathrm{Sym}(X) \cong S_3 \cong P\Gamma L_3(4)/L_3(4),$$

with the kernel $G_+^{(3)}$ being the element-wise stabilizer of X in G.

The whole of the Steiner system $(\mathcal{P}, \mathcal{B})$ can be described in terms of its residual projective plane Ξ as in the following lemma.

Lemma 6.6 *The following assertions hold:*

(i) $\mathcal{E}^{(3)}$ *are the lines of* Ξ*;*
(ii) $\mathcal{E}^{(2)}$ *are the hyperovals in* Ξ*;*
(iii) $\mathcal{E}^{(1)}$ *are the Fano subplanes in* Ξ*;*
(iv) $\mathcal{E}^{(0)}$ *are the sums of pairs of distinct lines in* Ξ*;*
(v) φ *is obvious on* $\mathcal{E}^{(3)}$ *and on* $\mathcal{E}^{(0)}$*; it is constant on* $G_+^{(3)}$*-orbits, thus, on labeling* X *by the orbits on the Fano subplanes, the map* φ *just assigns*

to $E \in \mathcal{E}^{(1)}$ *its* $G_+^{(3)}$*-orbit, and finally for* $E \in \mathcal{E}^{(2)}$ *the* 2*-set* $\varphi(E)$ *is the union of orbits of subplanes intersecting the hyperoval* E *in an odd number of points.*

Proof An easy combinatorial counting procedure gives the sizes of the $|\mathcal{E}^{(k)}|$ for $k = 3, 2, 1$ and 0 equal to 21, 168, 360 and 210, respectively. Item (i) has already been justified. If $E \in \mathcal{E}^{(2)}$ and $E_1 \in \mathcal{E}^{(3)}$ then $|\varphi(E) \cap \varphi(E_1)| = 2$ and hence $|E \cap E_1| \leq 2$ because of the intersection property, which implies that E_1 is a hyperoval. Since $|\mathcal{E}^{(2)}| = 168$ is the total number of hyperovals in Ξ, (ii) follows. The rest can be deduced from the fact that the sum of two octads intersecting in a 4-set is again an octad, together with Lemmas 3.16 and 3.17. □

The above lemma describes the unique way in which the Steiner system $(\mathcal{P}, \mathcal{B})$ can be recovered from its residual projective plane Ξ of order 4 and can easily be transferred into a uniqueness proof.[4]

6.4 Small Mathieu Groups

The maximal subgroup $G^{(4)}$ is the stabilizer of the partition of the point set $\mathcal{P}$ of the Steiner system into a pair of disjoint dodecads. This subgroup is the Mathieu group M_{12} extended by an automorphism of order 2. If we intersect $G^{(4)}$ with $G^{(1)}$ then we obtain the smallest Mathieu group M_{11}. Thus $G^{(4)}$ leads us to the 'small' Mathieu groups M_{11} and M_{12} which constitute a world of their own and thus deserve a few subsections. We have

$$|M_{12}| = 2^6 \cdot 3^3 \cdot 5 \cdot 11, \quad |M_{11}| = 2^4 \cdot 3^2 \cdot 5 \cdot 11.$$

6.4.1 Dodecads and M_{12}

Let $\mathcal{D}$ denote the set of dodecads in the Golay code $\mathcal{C}$, let D_1 and D_2 be a pair of disjoint dodecads and let $G^{(4)}$ be the stabilizer in G of the partition

$$\mathcal{P} = D_1 \cup D_2.$$

The set $\mathcal{D}$ of dodecads is the set of weight-12 codewords in the Golay code, and every dodecad can be represented as the sum of two octads intersecting in a pair. Since G acts transitively on the set of such pairs of octads (with the stabilizer of an ordered pair isomorphic to S_6), G acts transitively on $\mathcal{D}$ and

4 P. Dembowski, *Finite Geometries*, Springer, Berlin, 1968.

also on the set of complementary dodecads, while $G^{(4)}$ acts transitively on the set of pairs of octads whose sum is D_1 or D_2 and

$$|G^{(4)}| = \frac{|G| \cdot 2}{|\mathcal{D}|} = 12 \cdot 11 \cdot 10 \cdot 9 \cdot 8 \cdot 2.$$

Clearly $G^{(4)}$ contains an index-2 subgroup $G_+^{(4)}$ which is the stabilizer of the individual dodecads D_1 and D_2.

A significant amount of information on $G^{(4)}$ can be deduced by deciding how an octad can split between D_1 and D_2. Before proceeding to this topic let us state a characterization of the elements in G with large number of fixed points in $\mathcal{P}$ in the form of the following *fixed-points* property.

Lemma 6.7 *For $g \in G$ suppose that the set $f(g) = \{p \in \mathcal{P} \mid g(p) = p\}$ contains at least five elements. Then one of the following holds:*

(i) *g is the identity element;*
(ii) *$f(g)$ is an octad and g is of order 2 conjugate to an element in R_1;*
(iii) *$f(g)$ is a 6-set not contained in an octad, the order of g is 3, and $N_G(\langle g \rangle) \cong 3 \cdot S_6$ is a conjugate of a complement to R_3 in G_3.* □

Now let $B_1, B_2 \in \mathcal{B}$ with $A_0 := B_1 \cap B_2$ of size 2, so that $D_1 := B_1 + B_2$ and $D_2 := \mathcal{P} \setminus D_1$ are dodecads. If we set $A_1 = B_1 \setminus A_0$, $A_2 = B_2 \setminus A_0$ and $A_3 = D_2 \setminus A_0$ then

$$D_1 = A_1 \cup A_2 \text{ and } D_2 = A_0 \cup A_3,$$

where both unions are disjoint. We assume that B_1 is the octad stabilized by G_1, and we denote by π the ordered partition

$$\mathcal{P} = A_0 \cup A_1 \cup A_2 \cup A_3$$

and by Y the stabilizer of π in Y. By the *second complement* property Lemma 5.14 Y is an S_6-subgroup of G_1 that is not contained in an $L_4(2)$-complement to R_1. Furthermore, the members of the partition π are the orbits of Y on $\mathcal{P}$ with lengths 2, 6, 6 and 10. Clearly S_6 has a unique transitive representation of degree 2 (with kernel A_6) and a unique transitive representation of degree 10, which is the action by conjugation on the set of Sylow 3-subgroups. By Lemma 3.21 there are two representations of degree 6 which are conjugate in $\mathrm{Aut}(S_6)$ containing S_6 with index 2.

Lemma 6.8 *The actions of Y on A_1 and A_2 are not equivalent.*

Proof Let $t \in Y$ act on A_1 as a transposition, thus fixing four elements in A_1. Then the elements of A_3 can be viewed as partitions of A_1 into two 3-subsets, thus demonstrating that the number of elements in A_3 fixed by t is also four. By the fixed-points property Lemma 6.7 there are no further points fixed by t, in particular t acts on A_2 as the product of three commuting transpositions, proving the claim. □

Let $Y^{(0)}$ be the commutator subgroup of Y isomorphic to A_6 (which fixes the elements of A_0), let $Y^{(4)} = \mathrm{Aut}(Y)$, which is the stabilizer of π viewed as an unordered partition, thus permuting A_1 and A_2, and let $Y^{(1)}$, $Y^{(2)}$ and $Y^{(3)}$ be the proper subgroups of $Y^{(4)}$ properly containing $Y^{(0)}$. We assume that $Y^{(1)} = Y$ and that $Y^{(2)}$ is the subgroup which fixes the elements of A_0 and permutes A_1 and A_2.

A comparison of the number of dodecads with the number of pairs of octads with intersections of size 2 gives the following *dodecad pair-partition* principle.

Lemma 6.9 *For a pair $\{D_1, D_2\}$ of complementary dodecads and for a 2-subset A_0 of D_2 there is a unique disjoint partition $\mathcal{P} = A_0 \cup A_1 \cup A_2 \cup A_3$ such that $D_1 = A_1 \cup A_2$ with $A_1 \cup A_0, A_2 \cup A_0 \in \mathcal{B}$.* □

Let $\mathcal{F}$ be the set of all 6-subsets of D_1 which appear as A_1 or A_2 in a partition π as above. Then the following *dodecad Steiner* property holds.

Lemma 6.10 *The pair $(D_1, \mathcal{F})$ is a Steiner system of type $S(12, 6, 5)$.*

Proof By Lemma 6.9 we have

$$|\mathcal{F}| = \binom{12}{2} \cdot 2 = 132.$$

Since $(\mathcal{P}, \mathcal{B})$ is a Steiner system of type $S(24, 8, 5)$, every 5-subset T of D_1 is contained in a unique octad B. Since $B \cap D_1$ is an even subset which contains T and cannot be the whole of B, we have $|B \cap D_1| = 6$ and thus $B \cap D_1 \in \mathcal{F}$. □

Next we claim that the stabilizer $G_+^{(4)}$ of the set D_1 in G acts on D_1 5-fold transitively. In fact, $G_+^{(4)}$ acts transitively on the set $\mathcal{F}$ of blocks of the Steiner system on D_1 and the stabilizer $Y \cong S_6$ of $A_1 \in \mathcal{F}$ induces on A_1 the natural action of degree 6, which is 5-fold (and even 6-fold) transitive, and the claim follows by the *dodecad Steiner* property Lemma 6.10. Since

$$|G_+^{(4)}| = 12 \cdot 11 \cdot 10 \cdot 9 \cdot 8,$$

which is precisely the number of ordered 5-subsets of D_1, we have established the following theorem.

Theorem 6.11 *The group $G_+^{(4)}$ acts on D_1 sharply 5-fold transitively.* □

The group $G_+^{(4)}$ is the Mathieu group M_{12} discovered by É. Mathieu,[5] while $G^{(4)}$ is its automorphism group $\mathrm{Aut}(M_{12})$.

From the above consideration we have the following *octad–dodecad orbit lengths* attribute.

Lemma 6.12 *The group $G^{(4)} \cong M_{12}.2$ acts on the set $\mathcal{B}$ of octads with two orbits $\mathcal{B}_{2,6}^{(4)}$ and $\mathcal{B}_4^{(4)}$ whose lengths are 264 and 495, respectively. The orbit $\mathcal{B}_{2,6}^{(4)}$ consists of the octads B with*

$$\{|D_1 \cap B|, |D_2 \cap B|\} = \{2, 6\},$$

while $\mathcal{B}_4^{(4)}$ is formed by the octads with both the intersections having size 4. □

It is remarkable that $G^{(4)}$ acting on the set of trios also has a 495-orbit $\mathcal{T}_4^{(4)}$ and that the bipartite graph on $\mathcal{B}_4^{(4)} \cup \mathcal{T}_4^{(4)}$ with the adjacency relation via inclusion is a cubic graph on which $G^{(4)}$ acts edge-transitively (preserving the parts), realizing the largest Goldschmidt amalgam

$$G_5^1 = \{(Q_8 * Q_8).D_{12}, (Z_4 \times Z_4).S_3.D_8\}.$$

This means that the amalgam G_5^1 is a subamalgam in the amalgam $\{G_1, G_2\}$ from the Mathieu amalgam.

6.4.2 The Hadamard Matrix of Order 12

A *Hadamard matrix* $H_{(n)}$ of order n is an $n \times n$ matrix with entries ± 1 such that

$$H_{(n)}^{\mathrm{T}} H_{(n)} = H_{(n)} H_{(n)}^{\mathrm{T}} = n I_{(n)},$$

where $H_{(n)}^{\mathrm{T}}$ denotes the matrix transpose to $H_{(n)}$. Given a Hadamard matrix $H_{(n)}$ and a pair $\pi = (P, Q)$ of ± 1-monomial matrices, the matrix $P H_{(n)} Q$ is also a Hadamard matrix which is said to be *equivalent* to $H_{(n)}$. This means that a matrix which is equivalent to $H_{(n)}$ can be obtained by

[5] É. Mathieu, Mémoire sur l'étude des fonctions de plusieurs quantités sur la manière de les former et sur les substitutions qui les laissent invariable. *J. Math. Pure Appl.*, sér. II, **6** (1861), 241–328.

(1) permuting the rows of $H_{(n)}$;
(2) permuting the columns of $H_{(n)}$;
(3) negating some of the columns of $H_{(n)}$;
(4) negating some of the rows of $H_{(n)}$.

If $PH_{(n)}Q = H_{(n)}$ then $Q = H_{(n)}^{-1}P^{-1}H_{(n)}$ and the pair $\pi = (P, Q)$ is by definition an automorphism of $H_{(n)}$. The set of all automorphisms of $H_{(n)}$ forms a group under composition denoted by $A(H_{(n)})$. Clearly

$$Z(H_{(n)}) = \{(I_{(n)}, I_{(n)}), (-I_{(n)}, -I_{(n)})\}$$

is an order-2 subgroup in the centre of $A(H_{(n)})$. The transpose of $H_{(n)}$ is also a Hadamard matrix, and if $H_{(n)}$ and $H_{(n)}^{\mathrm{T}}$ are equivalent then $A(H_{(n)})$ can be extended by the transposing operation followed by suitable pre- and post-multiplications by monomial matrices.

Up to equivalence there is a unique Hadamard matrix of order 2, if the order n is more than 2 it has to be divisible by 4, and for large such n there are plenty of non-equivalent Hadamard matrices, although up to equivalence there are unique $H_{(4)}$, $H_{(8)}$ and $H_{(12)}$.

By applying the equivalence transformations (3) and (4) one can bring any Hadamard matrix to a *standard form* where the entries in the first column and the first row are all positive (thus equal to 1). By removing the first column and the first row in the standard form of a Hadamard matrix, and by changing -1s to 0s, one obtains the incidence matrix of a *Hadamard residue* design. The following result follows directly from definitions.

Lemma 6.13 *The Hadamard residue design $D_{(n)}$ is a 2-design with parameters*

$$v = 4l - 1, \ k = 2l, \ \lambda = l,$$

where $n = 4l$ is the order of $H_{(n)}$. □

The construction of a 2-design $D_{(n)}$ from a Hadamard matrix can be reversed in the obvious way, so that an order-n Hadamard matrix $H_{(n)}$ can be constructed given a 2-design with parameters as in Lemma 6.13.

Given a Hadamard matrix $H_{(n)}$ in a standard form, the stabilizer of the top left entry in the automorphism group consists of purely permutation matrices and their negatives, and coincides with

$$\mathrm{Aut}(D_{(n)}) \times Z(H_{(n)}),$$

where $D_{(n)}$ is the Hadamard residue design of $H_{(n)}$.

Table 6.1

	∞	0	1	2	3	4	5	6	7	8	9	10
∞	1	1	1	1	1	1	1	1	1	1	1	1
0	1	–	1	–	1	1	1	–	–	–	1	–
1	1	–	–	1	–	1	1	1	–	–	–	1
2	1	1	–	–	1	–	1	1	1	–	–	–
3	1	–	1	–	–	1	–	1	1	1	–	–
4	1	–	–	1	–	–	1	–	1	1	1	–
5	1	–	–	–	1	–	–	1	–	1	1	1
6	1	1	–	–	–	1	–	–	1	–	1	1
7	1	1	1	–	–	–	1	–	–	1	–	1
8	1	1	1	1	–	–	–	1	–	–	1	–
9	1	–	1	1	1	–	–	–	1	–	–	1
10	1	1	–	1	1	1	–	–	–	1	–	–

The design $D_{(4)}$ is a triangle and

$$A(H_{(4)}) \cong (Q_8 * Q_8) \colon S_3.$$

The design $D_{(8)}$ is the projective plane of order 2 and $A(H_{(8)})$ is the centralizer of a transvection in $L_5(2)$, so that

$$A(H_{(8)}) \cong 2_+^{1+6} \colon L_3(2).$$

The design $D_{(12)}$ is the quadratic residue design over $GF(11)$, and a standard form of $H_{(12)}$ is given in Table 6.1, where the rows and columns, except the first ones, are indexed by the elements of $GF(11)$ with the row with index $f \in GF(11)$ having positive entries precisely in the columns with indices in the set $\{\infty\} \cup (Q + f)$, where $Q = \{1, 3, 4, 5, 9\}$ is the set of non-zero squares in $GF(11)$.

In the rest of this section we provide a conceptual justification of the following result that was elaborated by M. Hall.[6]

Theorem 6.14 *The group $A(H_{(12)})$ acting on the sets of rows and columns of $H_{(12)}$ with kernel $Z(H_{(12)})$ realizes two non-isomorphic 5-transitive actions of degree 12 of the Mathieu group M_{12} of order $12 \cdot 11 \cdot 10 \cdot 9 \cdot 8$.*

If q is a prime power congruent to 3 modulo 4 then the incidence system whose points are the elements of $GF(q)$ and whose blocks are subsets of the form $Q + f$, where Q is the set of non-zero squares in $GF(q)$ and f runs through $GF(q)$, is a 2-design with parameters $v = q, k = (q-1)/2$ and $\lambda = (q-3)/4$.

[6] M. Hall, Note on the Mathieu group M_{12}, *Arch. Math.* **13** (1962), 334–340.

The semidirect product of the additive group of $GF(q)$ and the subgroup of the multiplicative subgroup of squares is a subgroup in the automorphism group of this *quadratic residue design*. Unless $q = 7$ or $q = 11$, this is the full automorphism group of the design.[7] When $q = 7$ the automorphism group has order 168, and in Section 3.4 this fact was essential to demonstrate the isomorphism $L_2(7) \cong L_3(2)$. In this chapter we take a look at the case $q = 11$ by proving the following lemma.

Lemma 6.15 *There exists a unique* 2*-design D with parameters $v = 11$, $k = 5$ and $\lambda = 2$. The automorphism group of D has order* 660 *and it is isomorphic to $L_2(11)$.*

Proof Let $D = (V, B)$ be a design with the parameters in the hypothesis of the lemma such that V is the set of points and B is the set of blocks. It follows immediately from the basic design theory that the dual design $D^* = (B, V)$ has the same parameters. Therefore the incidence graph Ξ of D is distance-regular with the following intersection diagram:

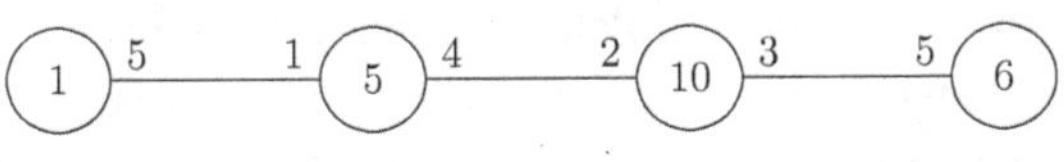

Since $\lambda = 2$, Ξ is a *rectograph* in the sense that it is a triangle-free graph in which any two vertices at distance 2 have precisely two common neighbours. Let x be a vertex of Ξ (which might be either a point or a block of D). As usual denote by $\Xi_i(x)$ the set of vertices at distance i from x in Ξ, with $\Xi_1(x)$ written simply as $\Xi(x)$.

The strategy is to describe the vertices of Ξ and their adjacencies in terms of certain 'structures' on $\Omega := \Xi(x)$ viewed as an abstract 5-set. Then x is a singleton, while $\Xi(x)$ is Ω by definition. By the rectograph property, $\Xi_2(x)$ is naturally indexed by the set $\binom{\Omega}{2}$ of 2-subsets of Ω, so that the adjacency between $\Xi(x)$ and $\Xi_2(x)$ is via inclusion. Let Π be the graph with the vertex set $V(\Pi) = \Xi_2(x)$ and edge-set

$$E(\Pi) = \{\{v, w\} \subseteq \Xi_2(x) \mid \Xi(u) \cap \Xi(w) \cap \Xi(x) = \emptyset\}.$$

Then Π brings the natural Petersen graph structure on $\Xi_2(x) = \binom{\Omega}{2}$. For $u \in \Xi_3(x)$ the set $\Xi(u)$ can be viewed as a 5-subset of vertices of Π. The set $\mathcal{X}$ of such subsets taken for all $u \in \Xi(u)$ has size 6 and satisfies the following statement.

[7] W. M. Kantor, Automorphism groups of Hadamard matrices, *J. Comb. Theory* **6** (1969), 279–281.

Lemma 6.16 (i) *every vertex of* Π *is in three subsets from* $\mathcal{X}$;
(ii) *every edge of* Π *is in two subsets from* $\mathcal{X}$;
(iii) *every non-edge of* Π *is in a unique subset from* $\mathcal{X}$. □

Returning to the proof of Lemma 6.15, we observe that by Lemma 6.16 (ii) the number of triples $(u, \{v, w\})$ with $u \in \Xi_3(x)$, $v, w \in \Xi_2(x) \cap \Xi(u)$ and $\{v, w\} \in E(\Pi)$ is

$$2\,|E(\Pi)| = 30 = 5\,|\Xi_3(x)|,$$

thus on average every subset from $\mathcal{X}$ contains five edges of Π. Since a shortest cycle in the Petersen graph is of length 5, every 5-subset of $V(\Pi)$ contains at most five edges and the bound is attained precisely when the subset is a 5-cycle. Thus $\mathcal{X}$ consists of the six 5-cycles in the Petersen graph Π. In total Π contains 12 cycles, each being stabilized by a dihedral subgroup D_{10} of $\mathrm{Sym}(\Omega) \cong S_5$. Under the action of $\mathrm{Alt}(\Omega)$ the set of 5-cycles splits into two orbits, each of size 6. Two cycles from different orbits intersect in zero or three vertices, while cycles from the same orbit always intersect in two vertices. By the above conditions (i) to (iii) in Lemma 6.16 $\mathcal{X}$ is an orbit of $\mathrm{Alt}(\Omega)$ on the set of 5-cycles in Π, which completes the uniqueness proof. It is clear now that Ξ is vertex-transitive with the stabilizer of a vertex isomorphic to A_5. In particular, D is isomorphic to its dual, and $\mathrm{Aut}(D)$ has order $|V| \cdot |A_5| = 660$.

In order to establish the isomorphism $\mathrm{Aut}(D) \cong L_2(11)$ we engage another model for D. It is well known[8] that $L_2(11)$ contains 22 subgroups isomorphic to A_5, which form two conjugacy classes, each of size 11, fused in

$$\mathrm{Aut}(L_2(11)) \cong PGL_2(11).$$

We take points to be one of these classes and blocks to be another one, with a point-subgroup incident to a block-subgroup whenever they intersect in a subgroup isomorphic to A_4. By calculating in the group $L_2(11)$, one can check that this construction does indeed lead to a 2-design with the required parameters. □

By Lemma 6.15 the stabilizer of the top left entry in the automorphism group of (the standard form of) $H_{(12)}$ is the group of order 1320 isomorphic to $L_2(11) \times 2$. On the other hand, every entry can be brought to the top left position by a suitable automorphism. In fact, given an (i, j)th position we first make the ith row and the jth column totally positive by negating suitable rows and columns, then for some suitable permutation matrices P and Q the matrix

[8] L. E. Dickson, *Linear Groups: With an Exposition of the Galois Field Theory*, Dover Publications, New York, 1958.

$PH_{(12)}Q$ is a Hadamard matrix with totally positive first row and first column. Because of the uniqueness of the 2-(11, 5, 2) design, we can make it equal to the original matrix $H_{(12)}$ by applying suitable permutation matrices P_1 and Q_1 to the top left position. Thus (P_1P, QQ_1) is an automorphism of $H_{(12)}$ which maps the (i, j)th position to the position (∞, ∞) and

$$|A(H_{(12)})| = 12 \cdot 12 \cdot 1320 = 12 \cdot 11 \cdot 10 \cdot 9 \cdot 8 \cdot 2.$$

In order to complete the proof of Hall's theorem, Theorem 6.14, first observe that the element-wise stabilizer of the leading five columns is trivial since only two rows have five equal leading positions and it is easy to check that the transposition of just these two rows does extend to an automorphism of $H_{(12)}$. Since the order of $A(H_{(12)})/Z(H_{(12)})$ is precisely the number of ordered 5-subsets in a 12-element set, the action of $A(H_{(12)})$ on columns is indeed sharply 5-fold transitive. A similar argument applies to rows, since $H_{(12)}$ is equivalent to its transpose. The stabilizer of a column induces the smallest Mathieu group M_{11} on the positions in that column and it does not stabilize a row. This can be seen, for example, by comparing the order of the stabilizer of a column with the order of the stabilizer of an entry. Therefore the action on the rows is on the cosets of a representative of another class of M_{11}-subgroups in $A(H_{(12)})/Z(H_{(12)})$, and these classes are fused in the extension by an automorphism of $A(H_{(12)})$ transposing $H_{(12)}$.

6.4.3 The Non-splitness of $A(H_{(12)})$

In view of the *octad–dodecad orbit lengths* attribute Lemma 6.12, given a Hadamard matrix $H_{(12)}$ we can construct a Steiner system $(\mathcal{P}, \mathcal{B})$ of type $S(24, 8, 5)$ in the following way:

(i) define $\mathcal{P}$ to be the union of the set $\mathcal{R}$ of rows and the set $\mathcal{K}$ of columns of $H_{(12)}$;

(ii) given a pair of distinct rows r_1 and r_2, define $E(r_1, r_2)$ and $D(r_1, r_2)$ to be the sets of columns where positions in these two rows are equal and different, respectively, and put

$$\mathcal{B}_{2,6}^{(r)} = \left\{ \{r_1, r_2\} \cup E(r_1, r_2), \{r_1, r_2\} \cup D(r_1, r_2) \mid \{r_1, r_2\} \in \binom{\mathcal{R}}{2} \right\};$$

(iii) define $\mathcal{B}_{2,6}^{(k)}$ with respect to the set $\mathcal{K}$ of columns according to a procedure similar to that in (ii);

(iv) given a quadruple $R = \{r_1, r_2, r_3, r_4\}$ of rows, define two subsets $K^{(\mathrm{o})}$ and $K^{(\mathrm{e})}$ of columns consisting of those having even and odd numbers of positive signs in the rows from R, choose $x(R) \in \{o, e\}$ such that $|K^{(x(R))}|$ is of size 4, and put

$$\mathcal{B}_4 = \left\{ R \cup K^{(x(R))} \mid R \in \binom{\mathcal{R}}{4} \right\};$$

(v) put $\mathcal{B} = \mathcal{B}_{2,6}^{(r)} \cup \mathcal{B}_{2,6}^{(k)} \cup \mathcal{B}_4$.

Then by Lemma 6.12 we have the following.

Lemma 6.17 *Let $H_{(12)}$ be representative of the unique equivalence class of Hadamard matrices of order* 12. *Then in the above terms* $(\mathcal{R} \cup \mathcal{K}, \mathcal{B}_{2,6}^{(r)} \cup \mathcal{B}_{2,6}^{(k)} \cup \mathcal{B}_4)$ *is a Steiner system of type* $S(24, 8, 5)$. □

The reverse construction of a Hadamard matrix from a Steiner system is not so straightforward, since it involves a cocycle, which leads to the non-splitness of $A(H_{(12)})$.

The construction proceeds as follows. We start with a Steiner system $(\mathcal{P}, \mathcal{B})$ of type $S(24, 8, 5)$, with $G \cong M_{24}$ being its automorphism group. Let

$$\mathcal{P} = D_1 \cup D_2$$

be an ordered partition of $\mathcal{P}$ into two disjoint dodecads, and let $G_+^{(4)} \cong M_{12}$ be the stabilizer of this partition in G. We form a matrix $H_{(12)}$ with rows indexed by the points in D_1 and with columns indexed by the points in D_2. Choose a point $\infty_1 \in D_1$ and a point $\infty_2 \in D_2$, and set the (∞_1, ∞_2)-entry in $H_{(12)}$ to be 1. By Lemma 6.12 there is a pair of $G_+^{(4)}$-invariant mappings

$$\chi_1 : \binom{D_1}{2} \to \binom{D_2}{6|6} \quad \text{and} \quad \chi_2 : \binom{D_2}{2} \to \binom{D_1}{6|6},$$

such that the image of χ_2 is the set of partitions of D_2 into pairs of complementary blocks of the Steiner system $(D_1, \mathcal{F})$ of type $S(12, 6, 5)$ on D_1 as in Lemma 6.10, and similarly for χ_1. Let χ_1^∞ be the restriction of χ_1 to the set of two-element subsets of D_1 containing ∞_1, where we modify the image by choosing from the pair of blocks the one which contains ∞_2. We define χ_2^∞ in a similar way.

Let $g \in G_+^{(4)}$, and let $P_0(g)$ and $Q_0(g)$ be the permutation matrices induced by the actions of g on D_1 and D_2, respectively. We shall define simultaneously the (ij)th entry in $H_{(12)}$ (where i and j are the images under g of ∞_1 and ∞_2, respectively) and the monomial matrices $P(g)$ and $Q(g)$ whose pair forms an automorphism of the Hadamard matrix to be constructed. If g stabilizes both ∞_1 and ∞_2 then $P(g) = P_0(g)$ and $Q(g) = Q_0(g)$. If $\infty_2^g \neq \infty_2$ then $P(g)$

is $P_0(g)$ multiplied from the right by the ± 1 diagonal matrix whose positive entries are precisely in the positions from

$$\chi_2^{\infty}(\{\infty_2, \infty_2^g\}).$$

The multiplier of $Q_0(g)$ is defined similarly. The result of the construction is summarized in the following statement.

Lemma 6.18 *For $g \in G_+^{(4)} \cong M_{12}$ define*

(i) $m_1(g) = 1$ *if either* $\infty_2^g = \infty_2$ *or* $\infty_1^g \in \chi_2^{\infty}(\{\infty_2, \infty_2^g\})$;
(ii) $m_1(g) = -1$, *otherwise.*

Let $m_2(g)$ be defined in a similar way. Then $m_1(g)m_2(g)$ is uniquely determined by the pair (∞_1^g, ∞_2^g) and is independent of the particular choice of g. Furthermore, the matrix $H_{(12)}$, whose (∞_1^g, ∞_2^g)-entry is equal to $m_1(g)m_2(g)$ is a Hadamard matrix and the pair $(P(g), Q(g))$ is an automorphism of $H_{(12)}$. □

The non-splitness follows from the following result.

Lemma 6.19 *For $i = 1$ and 2 let $\mathcal{D}_i$ be the $GF(2)$-permutation module of $G_+^{(4)} \cong M_{12}$ acting on D_i, let $\mathcal{E}_i$ be the co-dimension-1 submodule comprised by the even subsets of D_i, and let $\mathcal{H}_i = \mathcal{E}_i/\{\emptyset, D_i\}$ be the heart of the permutation module $\mathcal{D}_i$. For $\mathcal{X}_i$ being $\mathcal{D}_i$, $\mathcal{E}_i$ or $\mathcal{H}_i$ let $\mathcal{X}_i : M_{12}$ be the semidirect product of $\mathcal{X}_i$ and $G_+^{(4)} \cong M_{12}$ with respect to the natural action. Then the following assertions hold:*

(i) *$\mathcal{P}_i : M_{12}$ contains a single class of M_{12}-complements;*
(ii) *$\mathcal{E}_i : M_{12}$ contains two classes of M_{12}-complements;*
(iii) *$\mathcal{H}_i : M_{12}$ contains at least four classes of M_{12}-complements;*
(iv) *two classes of M_{12}-complements in* (iii) *are 'standard' in the sense that their pre-images in $\mathcal{E}_i : M_{12}$ are direct products $2 \times M_{12}$ and two classes are 'non-standard' with their pre-images being non-split extensions of the form $2 \cdot M_{12}$.*

Proof We start with (i) and suppose that $\mathcal{P}_i : M_{12}$ contains at least two classes of complements. Then by a standard theory, relating classes of complements with first cohomology and indecomposable extensions by trivial modules,[9] there exists an M_{12}-module $\widetilde{\mathcal{P}}_i$ containing $\mathcal{P}_i$ as a co-dimension-1 submodule but not as a direct summand. Considering the dual $\widetilde{\mathcal{P}}_i^*$ of $\widetilde{\mathcal{P}}_i$

[9] M. Aschbacher, *Finite Group Theory*, Cambridge University Press, Cambridge, 2000.

and making use of the fact that $\mathcal{P}_i$ is self-dual, we obtain an M_{12}-module homomorphism

$$\varepsilon : \widetilde{\mathcal{P}}_i^* \to \mathcal{P}_i$$

such that the kernel of ε is a trivial one-dimensional submodule which is not complemented by an M_{12}-submodule. Since $\mathcal{P}_i$ is the permutation module, it contains the natural basis set indexed by one-element subsets of D_i, which we also denote by D_i. The group M_{12} acts on this basis set in the natural 5-fold transitive way as on the cosets of M_{11}. Let $\widetilde{D}_i$ be the pre-image of D_i in $\widetilde{\mathcal{P}}_i^*$. Then

$$|\widetilde{D}_i| = |\ker(\varepsilon)| \cdot |D_i| = 24$$

and M_{12} acts either transitively on $\widetilde{D}_i$ or with two orbits, each of size 12. The first option is not possible, since M_{11} (being simple) does not contain subgroups of index 2, therefore $\widetilde{D}_i = \widetilde{D}_i^{(1)} \cup \widetilde{D}_i^{(2)}$ is a partition into M_{12}-orbits. Since the restrictions of ε to $\widetilde{D}_i^{(1)}$ and to $\widetilde{D}_i$ are both bijections onto D_i, and since D_i is a basis of $\mathcal{P}_i$, it is inevitable that $\widetilde{D}_i^{(1)}$ spans a copy of the permutation module isomorphic to $\mathcal{P}_i$ which complements the kernel of ε. This leads to a contradiction and hence proves (i). Notice that if M is an M_{12}-complement in $\mathcal{X}_i : M_{12}$ then

$$\mathcal{K}(\mathcal{X}_i, M) = \{M^x \mid x \in \mathcal{X}_i\}$$

is the conjugacy class of complements containing M. Therefore, the class of M in $\mathcal{P}_i : M_{12}$ splits into two classes determined by the parity of x. Hence (ii) follows from (i). To prove (iii) we make use of the isomorphism

$$\varphi : \mathcal{H}_1 : M_{12} \to \mathcal{H}_2 : M_{12},$$

which is implied by Lemma 6.12. Let M be an M_{12}-complement in $\mathcal{H}_1 : M_{12}$, and let x_1 and x_2 be odd subsets in D_1 and D_2, viewed as elements of $\mathcal{P}_1$ and $\mathcal{P}_2$, respectively. Then M^{x_1} and $(\varphi(M))^{x_2}$ are in different classes of M_{12}-complements in $\mathcal{H}_2 : M_{12}$. Notice that, since the kernel of the natural homomorphism $\mathcal{E}_i : M_{12} \to \mathcal{H}_i : M_{12}$ is the centre of the domain, the conjugations by x_1 and x_2 are well defined. Finally, (iv) is a direct consequence of (ii) and (iii). □

It can be seen from the construction of the Hadamard matrix $H_{(12)}$ from (D_1, D_2) before Lemma 6.18 that the matrices $P(g)$ for $g \in M_{12}$ project onto a non-standard complement in $\mathcal{H}_1 : M_{12}$ in the sense of Lemma 6.19 (iv) and we have the final result of the section.

Theorem 6.20 *The automorphism group $A(H_{(12)})$ of a Hadamard matrix of order* 12 *is a non-split extension* $2 \cdot M_{12}$. □

6.5 Turyn's Construction

A very elegant construction of the Golay code $\mathcal{C}$ is presented on pp. 49–50 of van Lint's book[10] and credited to R. J. Turyn. This can be viewed as a trio construction since the subgroup $L_3(2) \times S_3$ of M_{24} clearly visible from this construction is contained in the trio stabilizer $G^{(7)} = G_2 \cong 2^6 : (L_2(3) \times S_3)$, where the $L_3(2)$-direct factor is the subgroup $L^{(5)}$ in our notation.

Let $\mathcal{J}$ be the projective line over $GF(7)$ on which $L^{(5)}$ acts as $L_2(7)$, and let $\mathcal{J}_1, \mathcal{J}_2$ and $\mathcal{J}_3$ be disjoint eight-element sets. For every $1 \leq i \leq 3$ let us fix a bijection $\varphi_i : \mathcal{J} \to \mathcal{J}_i$ and turn $\mathcal{J}_i$ into an $L^{(5)}$-set by the rule

$$h \cdot \varphi_i(x) = \varphi_i(h(x))$$

for $h \in L^{(5)}$ and $x \in \mathcal{J}_i$. Let I_Q and I_N be the four-dimensional $L^{(5)}$-submodules of $2^{\mathcal{J}}$ as in Lemma 3.5. Let $\mathcal{P}$ be the disjoint union of $\mathcal{J}_1, \mathcal{J}_2$ and $\mathcal{J}_3$. Taking into consideration the column corresponding to $L^{(5)}$ in Table 5.1, we observe that the action of $L^{(5)}$ on the coordinates of the Golay code is similar to its action on the $\mathcal{P}$ we have just defined. If we put

$$\mathcal{C} = \{\varphi_1(X) + \varphi_1(Z) + \varphi_2(Y) + \varphi_2(Z) + \varphi_3(X) \\ + \varphi_3(Y) + \varphi_3(Z) \mid X, Y \in I_Q, Z \in I_N\},$$

then the following principle, known as *Turyn's* principle, holds.

Lemma 6.21 *The above-defined pair* $(\mathcal{P}, \mathcal{C})$ *is a Golay code.*

Proof We have to check the defining axioms. Since both I_Q and I_N are closed under the addition, so is $\mathcal{C}$. Since $(\mathcal{J}, I_Q)$ and $(\mathcal{J}, I_N)$ are isomorphic to the extended Hamming code at level 3 with weight enumerator

$$(0^1, 4^{14}, 8^1),$$

they are both doubly even and self-dual. Furthermore, $I_Q \cap I_N = \{\emptyset, \mathcal{J}\}$, which immediately implies that $\mathcal{C}$ is even. If X and Y run through a basis of I_Q and Z runs through a basis of I_N then the elements $\varphi_1(X) + \varphi_3(X)$, $\varphi_2(Y) + \varphi_3(Y)$ and $\varphi_1(Z) + \varphi_2(Z) + \varphi_3(Z)$ form a basis of $\mathcal{C}$, so that $\mathcal{C}$ is 12-dimensional. Since $\mathcal{C}$ is even, by the dimension consideration it must be self-dual. Since every element in the above basis of $\mathcal{C}$ we have just produced has size divisible

[10] J. H. van Lint, *Introduction to Coding Theory*, third edition, Springer, Berlin, 1999.

by 4, the same property holds for every element of $\mathcal{C}$. Finally, suppose that an element M of C contains fewer than eight elements. By the double evenness in this case M (if non-empty) contains just four elements. Since each of the three summands $\varphi_1(X)+\varphi_1(Z)$, $\varphi_2(Y)+\varphi_2(Z)$ and $\varphi_3(X)+\varphi_3(Y)+\varphi_3(Z)$ of M has an even size (because I_Q+I_N is contained in the even half of $2^{\mathcal{J}}$), we conclude that one of the summands must be the empty set. Since $I_Q \cap I_N$ is comprised by the empty set and the whole set $\mathcal{J}$, this leads us to the conclusion that Z is either empty or the whole of $\mathcal{J}$. Without loss of generality we assume that Z is empty. But since the subsets in I_Q have sizes 0, 4 and 8, it follows that M is empty, thus proving the minimal weight of $\mathcal{C}$ to be 8. □

It is easy to observe that $\mathcal{C}$ is stable under the symmetric group S_3 permuting $\mathcal{J}_i$s and φ_is accordingly, thus commuting with the action of $L^{(5)}$. It is less obvious from the above construction that $\mathcal{C}$ possesses the trio subgroup $2^6 : (L_3(2) \times S_3)$.

6.6 Quadratic Residues over $GF(23)$

The subgroup $G^{(8)} \cong L_2(23)$ preserves the structure of $\mathcal{C}$ as the quadratic residue over $GF(23)$. Consider the 24-set $\mathcal{P}$ as the projective line over $GF(23)$: this means that *the above-defined pair* $(\mathcal{P}, \mathcal{C})$ *is a Golay code*,

$$\mathcal{P} = GF(23) \cup \{\infty\},$$

on which $G^{(8)} \cong L_2(23)$ acts naturally. Let $N = \{-a^2 \mid a \in GF(23)\}$ be the set of non-squares in $GF(23)$, and let $\mathcal{N} = \{N + x \mid x \in GF(23)\}$ be the blocks of the corresponding designs of non-square residues. Then, as shown in Section 2.3,[11] the subspaces spanned by $\mathcal{N}$ together with the whole set $\mathcal{P}$ satisfy the axioms of the Golay code.

6.7 The Smallest Maximal Subgroup

In this section we discuss the smallest maximal subgroup $G^{(9)}$ of M_{24}, which is $L_2(7) \cong L_3(2)$ acting transitively on the 24-set $\mathcal{P}$. We start by considering $\mathcal{P}$ as a transitive $G^{(9)}$-set and provide an independent justification for its presence in M_{24} by reconstructing a $G^{(9)}$-invariant Golay code structure on $\mathcal{P}$, and applying the uniqueness of the Golay code $\mathcal{C}_{12}$. This construction involves the remarkable interference between $L_2(7)$- and $L_3(2)$-incarnations of $G^{(9)}$ highlighted in Chapter 2.

[11] A. A. Ivanov, *Geometry of Sporadic Groups* I, Cambridge University Press, Cambridge, 1999.

We start with an abstract group $G^{(9)} \cong L_2(7) \cong L_3(2)$ together with a 24-set $\mathcal{P}$ on which $G^{(9)}$ acts transitively. Since

$$|G^{(9)}|/|\mathcal{P}| = 7,$$

it is immediately evident that the action is on the cosets of a Sylow 7-subgroup. We would like to redefine this action in terms of some sort of conjugation. If S is a Sylow 7-subgroup of $G^{(9)}$ then $N_{G^{(9)}}(S) \cong F_{21}$ acting on S by conjugation has two orbits on the non-identity elements in S. If s is a generator of S then the orbits are s^Q and s^N, where $Q = \{1, 2, 4\}$ and $N = \{3, 5, 6\}$ are the non-zero squares and the non-squares in $GF(7)$. Thus $G^{(9)}$ has two classes of order-7 elements of size 24 each and $G^{(9)}$ acts on $\mathcal{P}$ as it acts by conjugation on one of these classes. The two classes are fused in $PGL_2(7)$, so we can choose either of them without loss of generality.

Next we are going to describe the structure of the $GF(2)$-permutation module $2^{\mathcal{P}}$ in order to grasp the possible location of the invariant Golay code $\mathcal{C}_{12}$. Since we are working with the $GF(2)$-spaces, it is natural to engage the $L_3(2)$ incarnation of $G^{(9)}$. The following result is well-known in modular representation theory.

Lemma 6.22 *The group $L_3(2)$, besides the trivial module V_1, possesses just three irreducible modules in characteristic 2 which are absolutely irreducible and are realizable over $GF(2)$: the natural module V_3, the dual natural module V_3^* and the Steinberg module V_8, which can be defined via the equality*

$$V_3 \otimes V_3^* = V_8 \oplus V_1. \qquad \square$$

Thus V_8 is the space of traceless 3×3 matrices over $GF(2)$ with respect to $L_3(2)$-action via conjugation. Consider $L_3(2)$ as the set of non-singular 3×3 matrices over $GF(2)$, and let s be an element of order 7 in that group. Then the characteristic polynomial $\chi(s)$ of (the 3×3 matrix realization of) s is irreducible, so either

$$\chi(s) = \lambda^3 + \lambda + 1 \text{ or } \chi(s) = \lambda^3 + \lambda^2 + 1.$$

Since the trace of a matrix is the second-highest term in its characteristic polynomial, the trace is zero for the former polynomial and non-zero for the latter. Furthermore, if $\lambda^3 + a\lambda^2 + b\lambda + c$ is the characteristic polynomial of s then the characteristic polynomial of s^{-1} is $c\lambda^3 + b\lambda^2 + a\lambda + 1$. On the other hand, s and s^2 have the same characteristic polynomial. Thus precisely half of the non-identity elements of S are realized by trace-zero matrices and, if s is such a matrix, then the other two are s^2 and s^4. Therefore exactly one class of elements of order 7 in $L_3(2)$ is realized by traceless matrices. These matrices can be viewed as vectors in the Steinberg module V_8.

By the above paragraph V_8 is a factor module of $2^{\mathcal{P}}$ when the latter is viewed as a $G^{(9)}$-module. Since the dimension of V_8 is equal to the order of a Sylow 2-subgroup of $G^{(9)}$, the Steinberg module is *projective* in the sense that it always splits as a direct summand. Thus V_8 is a submodule in $2^{\mathcal{P}}$. Furthermore, if s is a traceless element of order 7 in $L_3(2)$, then its stabilizer in $G^{(9)}$ is $S = \langle s \rangle$, which fixes in V_8 three vectors s, s^2 and s^4. Therefore, the space of $G^{(9)}$-invariant mappings of $2^{\mathcal{P}}$ onto V_8 (denoted by $\mathrm{Hom}_{G^{(9)}}(2^{\mathcal{P}}, V_8)$) is two-dimensional, which means that $2^{\mathcal{P}}$ contains three copies of V_8 which generate a 16-dimensional submodule. The remaining eight-dimensional contribution can be identified as follows. The projective line $\mathcal{J}$ over $GF(7)$ is a $G^{(9)}$-set with its element stabilizer being the normalizer of a Sylow 7-subgroup, which is a Frobenius group F_{21} of order 21. Thus there is a $G^{(9)}$-map of $\mathcal{P}$ onto $\mathcal{J}$ and hence the permutation module $2^{\mathcal{J}}$ is a factor-module of $2^{\mathcal{P}}$. Since the Steinberg module is projective, this gives the structure of $2^{\mathcal{P}}$ as a $G^{(9)}$-module.

Lemma 6.23 *The following equality holds:*

$$2^{\mathcal{P}} = V_8^{(1)} \oplus V_8^{(2)} \oplus 2^{\mathcal{J}},$$

where $V_8^{(1)}$ and $V_8^{(2)}$ are isomorphic to the Steinberg module V_8, and where the third Steinberg submodule of $2^{\mathcal{P}}$ denoted by $V_8^{(3)}$ is diagonal *to the above two.* □

Our ultimate goal is to find a $G^{(9)}$-invariant Golay code $\mathcal{C}_{12}$ inside $2^{\mathcal{P}}$. Let us first decide where to look for it in terms of the submodule structure of $2^{\mathcal{P}}$. Since $\mathcal{C}_{12}$ is 12-dimensional, it must contain one of the three Steinberg modules and a four-dimensional submodule from $2^{\mathcal{J}}$. By the structure of the permutation module in Section 3.5, there are just two candidates for the latter submodule, namely I_Q and I_N, where Q and N are the non-zero squares and the non-squares in $GF(7)$. It is reasonable to ask which of the six possible choices to make, and the following result demonstrates that the choice is irrelevant.

Lemma 6.24 *There are precisely six 12-dimensional $G^{(9)}$-submodules in $2^{\mathcal{P}}$, namely*

$$V_8^{(i)} \oplus I_A,$$

where $i = 1, 2, 3$ *and* $A = Q, N$. *They are regularly permuted by*

$$M := N_{\mathrm{Sym}(\mathcal{P})}(G^{(9)})/G^{(9)} \cong 3 \times 2.$$

Proof Consider the direct product $X = PGL_2(7) \times 3$. The normalizer of a Sylow 7-subgroup in X is $F_{42} \times 3$. Take an F_{42}-subgroup Y in that normalizer which is not in the $PGL_2(7)$-direct factor of X and consider the action of X on the set X/Y of the cosets of Y. This action is transitive of degree 24 and its restriction to $L_2(7)$ is the action on the cosets of a subgroup $L_2(7) \cap Y$ which is of order 7. Thus the action of $L_2(7)$ on X/Y and that on $\mathcal{P}$ are isomorphic. Since $L_2(7)$ is a normal subgroup in X with index 6, we achieve the lower bound for M. To obtain the upper bound it suffices to use the equality $PGL_2(7) = \mathrm{Aut}(L_2(7))$ and the fact that the centralizer of a transitive permutation action is semi-regular. Since V_8 is absolutely irreducible and since the action of $L_2(7)$ on $\mathcal{J}$ is primitive of degree 8, it can easily be seen that an element of order 3 in M permutes transitively the Steinberg submodules in $2^{\mathcal{P}}$ centralizing $2^{\mathcal{J}}$, while an element of order 2 in M centralizes each Steinberg submodule, turning $2^{\mathcal{J}}$ into the permutation module of $PGL_2(7)$ acting naturally on $\mathcal{J}$. Therefore, $X \cong PGL_2(7) \times 3$ is the normalizer of $G^{(9)}$ in $\mathrm{Sym}(\mathcal{P})$. □

By Lemma 6.24 either each of the six submodules is a Golay code or none of them is.

Before proceeding further let us introduce an extra piece of notation. Let

$$\varphi_i : V_8 \to V_8^{(i)}, \quad \varphi_i : v \mapsto v_i$$

be the $G^{(9)}$-isomorphisms, uniquely determined for each $i = 1, 2, 3$, so that $\varphi_{ij} : v_i \mapsto v_j$ is the isomorphism of $V_8^{(i)}$ onto $V_8^{(j)}$. In these terms we see an additional automorphism ψ of $2^{\mathcal{P}}$ (viewed as an abstract $G^{(9)}$-module), which swaps v_1 and v_2 for all $v \in V_8$ and commutes with $V_8^{(3)} \oplus 2^{\mathcal{J}}$. Notice that ψ is not an automorphism of $2^{\mathcal{P}}$ as a permutation module, so that there are at least two permutation module structures around.

In order to recover the permutation module structure on $V_8^{(1)} \oplus V_8^{(2)} \oplus 2^{\mathcal{J}}$ we need to find a basis $\mathcal{P}$ that is transitively permuted by $G^{(9)}$. Let s be a trace-zero matrix of order 7 in V_8 and let S be the Sylow 7-subgroup in $G^{(9)}$ generated by s. The subgroup S itself can be viewed as the non-zero vector from $2^{\mathcal{J}}$, which S stabilizes. Let $v(s)$ be a vector in $V_8^{(1)} \oplus V_8^{(2)} \oplus 2^{\mathcal{J}}$ stabilized by S such that the images of $v(s)$ under $L_2(7)$ form a basis set. By the above reasoning the projection of $v(s)$ on $2^{\mathcal{J}}$ must be S. The projection of $v(s)$ onto $V_8^{(1)}$ is s_1^i for $i = 1, 2, 4$ and, in view of the $N_{L_2(7)}(S)$ action on S, we can assume that $i = 1$.

Then the projection of $v(s)$ onto $V_8^{(2)}$ cannot be s_2, since $s_1+s_2 = s_3$ is in $V_8^{(3)}$ and $v(s)$ would be inside the submodule $V_8^{(3)} \oplus 2^{\mathcal{J}}$. Therefore, the projection of $v(s)$ onto $V_8^{(2)}$ is either s_2^2 or s_2^4. Taking into consideration the automorphism ψ from the previous paragraph, we assume without loss of generality that

$$v(s) = s_1 + s_2^2 + S$$

and

$$\mathcal{P} = \{s_1 + s_2 + \langle s\rangle \mid s \in V_8, |\langle s\rangle| = 7\}.$$

Now can start looking for the Golay code $\mathcal{C}_{12}$ inside $2^{\mathcal{P}}$. A useful hint is that the Sylow 7-subgroup in M_{24} has order 7 and it stabilizes an octad. Thus we should analyze 8-subsets in $\mathcal{P}$ stabilized by a Sylow 7-subgroup from $L_2(7)$ as possible octads. Let s and t be two elements of order 7 from $\mathcal{P}$ which generate different Sylow 7-subgroups S and T, respectively. Then S acts on $\mathcal{P}$ with three fixed points s, s^2 and s^4 and three orbits of length 7, which are

$$(t^j)^S := \{s^{-k}t^j s^k \mid 0 \le k \le 6\}$$

for $j = 1, 2, 4$. Therefore altogether S has nine invariant 8-subsets in $\mathcal{P}$:

$$O(i, j) := \{s^i\} \cup (t^j)^S,$$

for $i, j = 1, 2, 4$. There is an element r of order three in $L_2(7)$ such that

$$r^{-1}sr = s^2, \quad r^{-1}tr = t^4$$

and an element σ of order 2 which swaps s and t, inverting r. Taking this into account, the nine invariant 8-sets split into three orbits, each of size 3 and $\{O(1, j) \mid j = 1, 2, 4\}$ is a transversal on these orbits. In order to locate the corresponding codewords of weight 8 inside the permutation module $2^{\mathcal{P}}$, we require the following intrinsic property of the Steinberg module. Notice that the Killing form $\mathbf{tr}(v \cdot u)$, where $\cdot$ stands for the matrix product, is the only non-zero invariant bilinear form on V_8.

Lemma 6.25 *Let s and t be two trace-zero matrices in V_8 which generate in $L_3(2)$ different Sylow 7-subgroups S and T, respectively. Then s and t can be chosen to be perpendicular, in which case t^2 and t^4 are not perpendicular to s. Then the vector*

$$\Sigma_j := \sum_{u \in (t^j)^S} u$$

is equal to s, s^4 and s^2 for $j = 1, 2$ and 4, respectively.

Proof First notice that the Σ_js are S-invariant, and, since $(s^i, t^j) = 0$ only if $j = i^2$, it is clear that the Σ_js are non-zero. On the other hand, s, s^2 and s^4 are the only non-zero S-invariant vectors in V_8, with s being perpendicular to itself, but not to s^2 or s^4. Thus Σ_1 is an S-invariant non-zero vector in V_8 that is perpendicular to s, so it must be s itself. Now, to complete the justification, it suffices to conjugate by r.

Let $B(i, j)$ denote the sum of the vectors $v(u)$ taken for all $u \in O(i, j)$, where $O(i, j)$ is the above-defined 8-orbit orbit of S on $\mathcal{P}$. Then by the *double-seven* principle together with the equality $s + s^2 + s^4 = 0$, both of which hold in V_8, we immediately obtain

$$B(1, 1) = s_2 + \mathfrak{J}, \quad B(1, 2) = s_1^2 + \mathfrak{J} \text{ and } B(1, 4) = s_1^4 + s_2^4 + \mathfrak{J},$$

which are contained in the submodules of $V_8^{(2)}$, $V_8^{(1)}$ and $V_8^{(3)}$, respectively, directly summed with the trivial one-dimensional submodule of $2^{\mathfrak{J}}$. This is consistent with our expectation that $B(1, j)$ will be an octad in the invariant Golay code we are looking for. Thus we can stick with $O(1, 1)$, understanding its implicit dependence on s, take the 24 realizations, one for every $s \in \mathcal{P}$, and analyze the intersections to check whether they are consistent with the Golay code axioms. Notice that the 24 realizations of $B(1, 1)$ clearly span the direct sum of $V_8^{(2)}$ with the trivial submodule of $\mathfrak{J}$, so that the span is nine-dimensional, and we denote it by $\mathfrak{C}_9$. □

In order to accomplish the construction we switch to the $L_2(7)$ incarnation of $G^{(9)}$ and redefine its action on $\mathcal{P}$. Let $V_2(7)^*$ be the set of non-zero vectors in a two-dimensional $GF(7)$-space. Then $V_2(7)^*$ contains $7^2 - 1 = 48$ vectors and $SL_2(7)$ acts naturally on this set, with the centre acting as ± 1 scalar matrices and with the stabilizer of a vector being of order 7. Thus $\mathcal{P}$ can be viewed as $V_2(7)^*$, where a vector is identified with its negative. The following result is well known and easy to check directly.

Lemma 6.26 *The group $SL_2(7) = Sp_2(7)$, when acting on the two-dimensional $GF(7)$-space preserves the symplectic form*

$$[x, y] = x_1 y_2 - x_2 y_1,$$

where $x = (x_1, x_2)$ and $y = (y_1, y_2)$ are written in the standard basis. □

Since the multiplicative group of $GF(7)$ is the direct product of the order-2 subgroup ± 1 and the subgroup $Q = \{1, 2, 4\}$ of non-zero squares, the symplectic form enables us to define an invariant mapping on $\mathcal{P} \times \mathcal{P}$ with values in $\{0\} \cup Q$ with $[p, q]$ being the value of the symplectic form of suitable representatives of p and q in $V_2(7)$ which lies in the above domain.

For $j \in Q$ define a graph $\Upsilon^{(j)}$, where $p, q \in \mathcal{P}$ are adjacent whenever $[p, q] = j$. The graph is known as the *Klein graph*,[12] and it is distance-regular with the following intersection array:

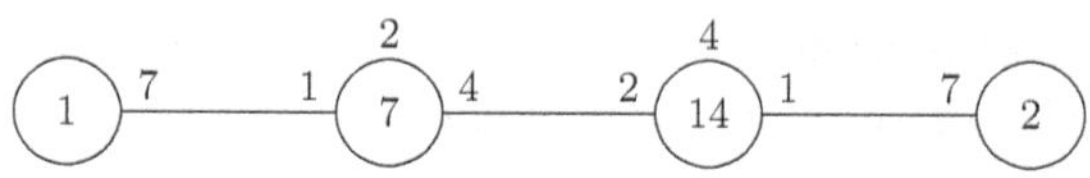

Furthermore, the isomorphism type of the graph is independent of the choice of $j \in Q$. The group $L_2(7)$ acts on $\Upsilon^{(j)}$ vertex- and edge-transitively. Under the stabilizer of a vertex (this being a cyclic group of order 7) the vertices at distances 2 and 3 split into two orbits each, while the action of $PGL_2(7)$ is distance-transitive. This is immediately evident from consideration of the normalizer of $L_2(7)$ in the symmetric group of $\mathcal{P}$, which we already know about in detail. In fact $PGL_2(7)$ is the full automorphism group of $\Upsilon^{(1)}$, but we do not require this information.

By the previous paragraph, without loss of generality we deal with the graph $\Upsilon = \Upsilon^{(1)}$. The graph is known to be a *local heptagon*, and we are going to see this and some other features of the graph a few lines below. Let x be the vertex of Υ which is represented by the vector $(0, 1)$. Then the seven neighbours of x are represented by the vectors $(1, a)$ taken for all $a \in GF(7)$ and we denote the vertex represented by $(1, a)$ simply by a. Then a and b from $\Upsilon(x)$ are adjacent if and only if

$$a - b = \pm 1 \bmod 7.$$

Therefore the subgraph induced by $\Upsilon(x)$ is indeed a heptagon. If y is a vertex at distance 2 from x in Υ then y is adjacent to precisely two vertices, say e and d from $\Upsilon(x)$, and, depending on the orbit of $G^{(9)}(x)$ containing y, the difference $e - d$ is equal to ± 2 or ± 4 modulo 7. Finally, the two vertices at distance 3 from x are represented by the vectors $(2, 0)$ and $(4, 0)$, respectively.

Now the set $\{x\} \cup \Upsilon(x)$ is a subset of size 8 that is invariant under the subgroup S of order 7 in $L_2(7)$ which is the stabilizer of x, and in view of the symmetry under $N_{\mathrm{Sym}(\mathcal{P})}(G^{(9)})$ we can assume that

$$\{x\} \cup \Upsilon(x) = O(1, 1),$$

which means that $\mathcal{C}_9$ is spanned by the 24 subsets of size 8 in $\mathcal{P}$, where each subset is a vertex of Υ together with its seven neighbours. Notice that, staying

12 A. E. Brouwer, A. M. Cohen and A. Neumeier, *Distance Regular Graphs*, Springer, Berlin, 1989, p. 386.

within the $L_2(7)$ incarnation, in order to be able to conclude that the dimension of $\mathcal{C}_9$ is indeed 9, one should have determined the $GF(2)$-rank of a 24×24 matrix. On the other hand, the setting in terms of the Klein graph enables one to get hold of intersections of the spanning 8-subsets and more, which is contained in the following *Klein* principle.

Lemma 6.27 *For Υ being the Klein graph and for x being a vertex of Υ, denote by x^* the union of $\{x\}$ with its neighbours: $x^* = \{x\} \cup \Upsilon(x)$. Then, for any two vertices x and y of Υ, the intersection $\iota(x, y) := x^* \cap y^*$ always contains an even number of vertices. However, for every four-vertex subset R there is a vertex z of Υ such that $R \cap z^*$ contains an odd number of vertices.*

Proof It is clear that the number of vertices in $\iota(x, y)$ is 8, 4, 2 and 0 when the distance from x to y is 0, 1, 2 and 3, respectively, which proves the first assertion. In order to prove the second part of the statement, let R be a set of four vertices in Υ. In order to have an even intersection of R with z^* for $z \in R$, the number of vertices in R adjacent to z must be odd. Since Υ does not contain complete subgraphs on four vertices, at least one vertex $z \in R$ is adjacent to just one other vertex in R. That vertex z is therefore adjacent to six vertices outside R. Since there are three vertices in R other than z, with one of them adjacent to z, any two vertices in Υ will have one or two common neighbours, depending on whether they are adjacent or not. Since $6 - (2 \cdot 2 + 1) = 1$, there is at least one vertex outside R, whose only neighbour in R is z, which completes the proof. □

Let (,) denote the symplectic form on $2^{\mathcal{P}}$ with respect to which $\mathcal{P}$ is an orthonormal basis and let q be the $GF(2)$-valued map on the set of even subsets from $2^{\mathcal{P}}$ defined by

$$q(A) = |A|/2 \bmod 2.$$

Since $|\mathcal{P}|$ is divisible by 4, q is a well-defined quadratic form associated with the form (,). By the first claim of the Klein principle Lemma 6.27, the code $\mathcal{C}_9$ is totally singular with respect to (,), and, since it is spanned by eight-element subsets, the quadratic form vanishes on $\mathcal{C}_9$. This also saves us from needing to carry out explicit calculations to recover (,) on $2^{\mathcal{P}}$, which is an abstract setting of the latter: the form vanishes on $V_8^{(1)}$ and on $V_8^{(2)}$ with $(v^{(1)}, u^{(2)}) = \mathbf{tr}(v \cdot u)$, $2^{\mathcal{J}}$ is perpendicular to $V_8^{(1)} \oplus V_8^{(2)}$ and with respect to the restriction of (,) to $2^{\mathcal{J}}$ the set $\mathcal{J}$ is an orthonormal basis. Then

$$\mathcal{C}_9^{\perp} = V_8^{(2)} \oplus 2^{\mathcal{J}}.$$

Theorem 6.28 *The code $V_8^{(2)} \oplus I_Q$ is a Golay code $\mathcal{C}_{12}$.*

Proof Since $I_Q^{\perp} = I_Q$ is in $2^{\mathfrak{d}}$, we conclude that $\mathcal{C}_{12}$ is a maximal totally singular subspace with respect to $(\ , \ )$. We claim that $\mathcal{C}_{12}$ is doubly even in the sense that the size of every subset in $\mathcal{C}_{12}$ is divisible by 4. This is equivalent to the claim that the quadratic form q vanishes on $\mathcal{C}_{12}$. But the latter claim is clear since $(\ , \ )$ vanishes on $\mathcal{C}_{12}$, while $\mathcal{C}_{12}$ is spanned by subsets of sizes divisible by 4. Finally, by the second part of the Klein principle Lemma 6.27, there are no four-element subsets in $\mathcal{C}_{12}$, so that the minimal weight is 8 and $\mathcal{C}_{12}$ is indeed the Golay code. □

7

The 45-Representation of M_{24}

The group M_{24} possesses a pair of algebraically conjugate irreducible representations of dimension 45. These representations are special in many respects. (1) When G. Frobenius[1] calculated the character table of M_{24} he deduced the degree-45 characters at the very end from the orthogonality relations when all other irreducibles had been obtained by an induction–restriction procedure starting from the permutation characters of small degrees (this is because the 45-dimensional M_{24}-module does not possess vectors with 'large' stabilizers). (2) The minimal degree of an irreducible character of the fourth Janko group, J_4, is 1333 and the character restricts to the largest subgroup $2^{11} : M_{24}$ as $1333 = 45 + 1288$. Because of this, the 45-dimensional representation of M_{24} played an essential role in the computer-free construction of J_4.[2] (3) According to the epoch-making Mathieu Moonshine observation,[3] the pair of 45-dimensional representations of M_{24} corresponds to the linear term in the massive sector of the elliptic genus of $K3$. (4) Thompson's uniqueness criterion for the Mathieu amalgam is satisfied for the 45-dimensional module, thus this is a very natural framework in which to represent the amalgam (see the end of Section 1.7). In this section we demonstrate how a 45-dimensional representation of M_{24} can be constructed within the amalgam setting.

7.1 Representing $L_3(2)$ in Three Dimensions

Let $L \cong L_3(2)$, let k be a field of non-even characteristic which contains a square root of -7, and let $W_3(k)$ be a three-dimensional k-space. It has

[1] G. Frobenius, Über die Charaktere der mehrfach transitiven Gruppen, *Sitzungsber. Königl. Preuß. Akad. Berlin* (1904), 558–571.

[2] A. A. Ivanov and U. Meierfrankenfeld, A computer-free construction of J_4, *J. Algebra* **219** (1999), 113–172.

[3] T. Eguchi, H. Ooguri and Y. Tachikawa. Notes on the $K3$ surface and the Mathieu group M_{24}, *Exp. Math.* **20** (2011), 91–96.

been known ever since the work of Felix Klein that L possesses a pair of algebraically conjugate irreducible representations

$$\varphi\colon L \to GL(W_3(k)).$$

The explicit form of φ and the distinction between algebraic conjugates is not relevant. What we require is the following S_4-*restriction* principle.

Lemma 7.1 *The restriction of the three-dimensional representation $\varphi\colon L \to GL(W_3(k))$ of $L \cong L_3(2)$ to an S_4-subgroup is irreducible, being induced from a linear representation of a D_8-subgroup with a cyclic kernel. In particular, the character value on every involution is -1.*

Proof Consider the normal Klein 4-subgroup K_4 of S_4. Since K_4 contains three subgroups of index 2 which are conjugate in S_4 and 3 is the dimension of $W_3(k)$, we conclude that there is a K_4-eigenspace decomposition

$$W_3(k) = W_1 \oplus W_2 \oplus W_3.$$

For the sake of explicitness we assume that the kernels at W_1, W_2 and W_3 are generated by elements $(1\,2)(3\,4)$, $(1\,3)(2\,4)$ and $(1\,4)(2\,3)$, respectively. If $t = (1\,2)$ then t stabilizes W_1 and swaps W_2 and W_3. Since t has order 2, we can choose a basis (v_1, v_2, v_3) in $W_3(k)$ with $v_i \in W_i$, such that

$$t = (\varepsilon v_1)(v_2, v_3),$$

where ε is ± 1. Since L is simple, the image of φ is contained in $SL(W_3(k))$, so that the determinant of t on $W_3(k)$ is 1, which forces ε to be -1. On the other hand, the element $s = (1\,3\,2\,4)$ of order 4 acts as

$$s = (v_1)(v_2, -v_3, -v_2, v_3),$$

so that s is in the kernel of the action on W_1 of

$$C_{S_4}((1\,2)(3\,4)) \cong D_8.$$

Since the index of D_8 in S_4 is 3, which is the same as the dimension of $W_3(k)$), the restriction of φ to S_4 is induced from the action of the above D_8-subgroup on W_1. At this point the character on the involution is easily seen to be -1. □

7.2 An Explicit Form of $L_3(2) \to L_3(\mathbb{Q}(\sqrt{-7}))$

For the sake of completeness we demonstrate in this section how to produce an explicit form of a three-dimensional representation of $L \cong L_3(2)$ over the field of rational numbers extended by a square root of -7.

We consider L as the direct factor of the Levi complement of H_2, so that L is the group of non-singular linear transformations of a three-dimensional $GF(2)$-space U_3 with the basis (b_1, b_2, b_3). Along with the familiar transvections τ_{ij} for $1 \leq j, i \leq 3$, let us introduce the elements σ_{kl} of L, which swap the basis vectors b_k and b_l, preserving the remaining basis vector. If we put

$$E = \langle \tau_{21}, \tau_{31}, \tau_{32} \rangle, \quad F_1 = \langle E, \sigma_{12} \rangle, \quad F_2 = \langle E, \sigma_{23} \rangle,$$

then $E = C_L(\tau_{31}) \cong D_8$ is a Sylow 2-subgroup of L, and F_1 and F_2 constitute a pair of S_4-subgroups in L, being the stabilizers of the 1-subspace $\langle b_3 \rangle$ and of the 2-subspace $\langle b_2, b_3 \rangle$, respectively, with the Klein 4-subgroups in these S_4s being

$$K_1 = \langle \tau_{31}, \tau_{32} \rangle \text{ and } K_2 = \langle \tau_{31}, \tau_{21} \rangle,$$

respectively.

Let k be an extension of the field $\mathbb{Q}$ of rational numbers (the particular form of k will emerge later), let $W = W_3(k)$ be a three-dimensional k-space and let

$$\lambda_1 : F_1 \to GL(W)$$

be the representation of F_1 induced from the linear representation of E whose kernel is the cyclic subgroup of order 4 (this choice is motivated by the S_4-*restriction* principle Lemma 7.1). Let

$$W = W_1 \oplus W_2 \oplus W_3$$

be the decomposition of W with respect to K_1, such that τ_{31} centralizes W_1, τ_{32} centralizes W_2 and $\tau_{31}\tau_{32}$ centralizes W_3. Then $\lambda_1(F_1)$ consists of the ± 1 monomial matrices with determinant 1 in a suitable basis (v_1, v_2, v_3), where v_i spans W_i for $1 \leq i \leq 3$. Since τ_{21} acting by conjugation swaps τ_{32} and $\tau_{31}\tau_{31}$, centralizing τ_{31}, while σ_{12} swaps τ_{31} and τ_{32}, we have

$$\lambda_1(\tau_{21}) = (v_1, -v_1)(v_2, v_3), \quad \lambda_1(\sigma_{12}) = (v_1, v_2)(v_3, -v_3).$$

To diagonalize the action of K_2 we change the basis (v_1, v_2, v_3) of W to the basis (u_1, u_2, u_2), where

$$u_1 = v_1, \quad u_2 = \frac{1}{2}(v_2 + v_3), \quad u_3 = \frac{1}{2}(v_2 - v_3).$$

Then, in the new basis, K_2 acts diagonally and this action can be extended to a representation λ_2 of F_2 that is also induced from a linear representation of E with a cyclic kernel of order 4. Under λ_2 the form of τ_{31} is unchanged, while τ_{21} acts as τ_{32} acts in the old basis. On the other hand, for σ_{23} we have

$$\lambda_2(\sigma_{23}) = (u_1, -u_1)(u_2, u_3).$$

At this stage we have constructed a pair of representations λ_1 and λ_2 of F_1 and F_2 in $GL(W)$, which coincide when restricted to the intersection $F_1 \cap F_2 = E$. But the images of these representations do not immediately generate L, since in L the product of σ_{12} and σ_{23} has order 3. In order to satisfy this condition we 'shift' the representation λ_1 while preserving its restriction to E. This can be achieved by conjugating σ_{12} by an element which multiplies v_1 by a non-zero element x of k, preserving the other two vectors of the old basis (v_1, v_2, v_3). The shifted σ_{12} is represented by the matrix

$$\lambda_1^{(x)}(\sigma_{12}) = \begin{pmatrix} 0 & x & 0 \\ x^{-1} & 0 & 0 \\ 0 & 0 & -1 \end{pmatrix}.$$

We keep λ_2 unchanged and $\lambda_2(\sigma_{23})$ in the basis (v_1, v_2, v_3) has the following form:

$$\lambda_2^{(x)}(\sigma_{23}) = \begin{pmatrix} 0 & 1 & 1 \\ \frac{1}{2} & -\frac{1}{2} & \frac{1}{2} \\ \frac{1}{2} & \frac{1}{2} & -\frac{1}{2} \end{pmatrix}.$$

The condition $(\lambda_1^{(x)}(\sigma_{12})\lambda_2(\sigma_{23}))^3 = 1$ leads to the equation $x^2 + \frac{1}{2}x + \frac{1}{2} = 0$, whose roots are

$$x_{1,2} = \frac{-1 \pm \sqrt{-7}}{4}.$$

Taking either of these values for x does indeed lead to a representation of $L \cong L_3(2)$ as on pp. 310–311 of Burnside's book.[4] We conclude this section by presenting the following lemma.

Lemma 7.2 *The pairs* $(\lambda_1^{(x)}, \lambda_2)$ *describe all completions of the amalgam* (F_1, F_2) *in* $GL_3(k)$ *subject to the* S_4-restriction *principle Lemma* 7.1, *while the pairs* $(\lambda_1^{(x_i)}, \lambda_2)$ *for* $i = 1$ *and* 2 *describe the representations of the* $L_3(2)$*-completion of this amalgam.*

Proof The result is immediately evident from Goldschmidt's lemma, since E acts on W with two irreducible components W_1 and $W_2 \oplus W_3$, while the actions of F_1 and F_2 are irreducible. Notice that

$$(F_1, F_2) \cong (S_4, S_4)$$

is the amalgam G_3 from Goldschmidt's list. If $x = x_i$ for $i = 1$ or 2 then the equality

[4] W. Burnside, *Theory of Groups of Finite Order*, second edition, Cambridge University Press, Cambridge, 1911.

$$(\lambda_1^{(x_i)}(\sigma_{12})\lambda_2(\sigma_{23}))^3 = 1$$

implies that the coset graph of the completion contains a cycle of length 6, which enables us to identify the coset graph with the incidence graph of the projective plane of order 2. □

7.3 Hyperplanes in R_2

In this section we describe the hyperplanes (that is, index-2 subgroups) in R_2, their orbits and their stabilizers. This is required for constructing representations of G_2. Since G_2 and H_2 are two conjugate subgroups in $\mathrm{Aut}(H'_{12})$, we start by calculating in H_2, which is a subgroup in $L_5(2)$, and then formulate our discoveries in G_2-terms by applying the conjugating automorphism α.

The subgroup $R_2 = O_2(H_2) = O_2(G_2)$ has order 2^6 and contains three submodules isomorphic to the natural module of

$$K_2^+ = L \cong H'_{12}/R_2 \cong L_3(2).$$

The submodules are $R_2^{(4)}$, $R_2^{(5)}$ and $R_2^{(4\frac{1}{2})}$, which are generated by the transvections τ_{4i}, τ_{5i} and $\tau_{4i}\tau_{5i}$, respectively (where i runs from 1 to 3). The three submodules are naturally permuted by

$$K_2^- \cong H_2/H'_{12} \cong S_3.$$

If U_3 is the subspace in the natural module V of $H \cong L_5(2)$ spanned by b_1, b_2 and b_3 then for $j = 4,\ 5$ and $4\frac{1}{2}$ there is an L-isomorphism

$$\psi_j : U_3 \to R_2^{(j)}$$

which sends b_i onto τ_{4i}, τ_{5i} and $\tau_{4i}\tau_{5i}$, respectively. Clearly

$$\psi_4(v) + \psi_5(v) + \psi_{4\frac{1}{2}}(v) = 0$$

for every $v \in U_3$.

Lemma 7.3 *The group H_2 acts on the 63 hyperplanes in R_2 with two orbits $\mathcal{P}_1$ and $\mathcal{P}_2$ such that*

(i) $|\mathcal{P}_1| = 21$, *if $P \in \mathcal{P}_1$ then $N_{H_2}(P)/R_2 \cong S_4 \times 2$ and P contains $R_2^{(j)}$ for $j = 4, 5$ or $4\frac{1}{2}$;*
(ii) $|\mathcal{P}_2| = 42$, *if $P \in \mathcal{P}_2$ then $N_{H_2}(P)/R_2 \cong S_4$ and P contains neither of the $R_2^{(j)}$s.*

Proof Clearly a hyperplane P in R_2 intersects every $R_2^{(j)}$ in a subspace of co-dimension at most 1. If one of these co-dimensions is zero then P is uniquely determined by that j together with the hypeplane in U_3 which maps bijectively onto $P \cap R_2^{(k)}$ for $k \neq j$. These hyperplanes constitute $\mathcal{P}_1$. Alternatively, the pre-images

$$l_j := \psi_j^{-1}(P \cap R_2^{(j)}), \quad \text{for } j = 4,\ 5,\ 4\tfrac{1}{2}$$

form the complete triple of lines in the projective plane of U_3 passing through a point p. □

7.4 Representing G_1

In this section we define and analyze a 45-dimensional k-representation θ_1 of G_1. We take

$$G_{12} \cong R_1 : (2^3 : L_3(2)) \cong R_2 : (L_3(2) \times 2)$$

and define its three-dimensional representation $\theta_{12}^{(3)}$ with kernel R_2, where the $L_3(2)$-direct factor of the quotient over R_2 acts via the representation φ as in Lemma 7.1 and with the order-2 direct factor acting via scalar ± 1 operators.

The representation θ_1 is $\theta_{12}^{(3)}$ induced from G_{12} to G_1. Let X_1 be the 45-dimensional k-space supporting θ_1 and let χ_1 denote the character of θ_1. When analysing the action of G_1 on X_1 we identify G_1 with H_1, which is the stabilizer of the 1-subspace $V_1 = \langle b_5 \rangle$ in the natural module V of $H = L_5(2)$. The following four features are direct implications of the definition of θ_1 and the basic properties of induced modules.

(i) If U is a 2-subspace of V containing V_1 then the corresponding hyperplane $P(U)$ in R_1 has three-dimensional centralizer $X_1(U)$ in X_1, in particular $\chi_1(r) = -3$ for every non-identity element of R_1.
(ii) The stabilizer of $X_1(U)$ in G_1 coincides with the stabilizer $G_1(U)$ of U with $G_1(U) \cong G_{12}$.
(iii) If $g \in G_1(U)$ acts trivially on V/U then it acts on $X_1(U)$ as a plus or minus scalar operator depending on the triviality or non-triviality of its action on U.
(iv) The $L_3(2)$-direct factor of $G_1(U)/C_{G_1}(V/U) \cong L_3(2) \times 2$ acts on $X_1(U)$ via the three-dimensional representation φ as in Lemma 7.1.

The above features enable us to calculate the values of the character χ_1 of θ_1 on the elements τ_{21} and $\tau_{21}\tau_{53}$. In a sense these character values tell us which amalgam we are going to represent: the $L_5(2)$-amalgam or the M_{24}-amalgam. This is because the involution τ_{21} is contained in the Levi complement

$L = K_2^+$ to R_2 in H'_{12}, and in H_2 it is centralized by an element of order 3 from K_2^-. On the other hand, $\tau_{21}\tau_{53}$ is contained in the complement $L^{(5)}$, which is centralized by an element of order 3 in G_2. In terms of Section 5.1 the actions of L and $L^{(5)}$ on the 24-element set $\mathcal{P}$ are also different: $L^{(5)}$ acts with three orbits of length 8 each, so that $\tau_{21}\tau_{53}$ acts on $\mathcal{P}$ fixed-point freely, whereas L acts with orbit lengths 8, 7, 7, 1 and 1, so that τ_{21} fixes eight points in $\mathcal{P}$ which are easily seen to form an octad. Therefore, in M_{24} the involution τ_{21} is central, conjugate to involutions in R_1, whereas $\tau_{21}\tau_{53}$ is non-central. The character values we are after are given in the following *character* attribute.

Lemma 7.4 *The following equalities hold:*

$$\chi_1(\tau_{21}) = -3, \quad \chi_1(\tau_{21}\tau_{53}) = 5,$$

so that the character value of τ_{21} is equal to that of an involution from R_1.

Proof The transvections τ_{21} and τ_{53} act in the basis $(b_1, b_2, \ldots, b_5)$ via

$$\tau_{21} : b_1 \mapsto b_1 + b_2, \quad \tau_{53} : b_3 \mapsto b_3 + b_5,$$

centralizing the other basis vectors. Since τ_{35} is contained in R_1, it stabilizes every 2-subspace U containing $V_1 = \langle b_5 \rangle$. If U is also stabilized by τ_{21}, it is spanned by b_5 together with a vector from $b_1^* \cap b_5^*$, so that there are seven such subspaces U. One of them, namely $U_0 := \langle b_2, b_5 \rangle$, is special since τ_{21} and $\tau_{21}\tau_{53}$ act trivially both on U_0 and on V/U_0. Thus U_0 contributes 3 to the values of $\chi_1(\tau_{21})$ and $\chi(\tau_{21}\tau_{53})$. If U is one of the remaining six subspaces, then τ_{21} centralizes U and acts on V/U as a transvection. Therefore such a U contributes -1 to the value of $\chi_1(\tau_{21})$. On the other hand, $\tau_{21}\tau_{53}$, while still acting on V/U as a transvection, centralizes U only if it is contained in b_3^*. Otherwise, say for $U = \langle b_3, b_5 \rangle$, the action on U is non-trivial and the contribution to $\chi_1(\tau_{21}\tau_{53})$ is the negative of that to $\chi_1(\tau_{21})$. Since $b_1^* \cap b_3^* \cap b_5^*$ contains just three non-zero vectors, including b_2, we obtain the following equalities:

$$\chi_1(\tau_{21}) = 3 - 6 = -3, \quad \chi_1(\tau_{21}\tau_{53}) = 3 + 4 - 2 = 5.$$

□

7.5 Restricting from G_1 to G_{12}

In this section we restrict the representation θ_1 from G_1 to G_{12} and determine the decomposition of X into eigenspaces of R_2. The restriction θ_{12} of θ to G_{12} is easily seen to be the direct sum of the original three-dimensional

Table 7.1

	τ_{51}	τ_{52}	τ_{53}	τ_{41}	τ_{42}	τ_{43}
$U^{(1)}$	1	1	2	1	1	$U^{(1)}$
$U^{(2)}$	1	1	2	1	1	$U^{(2)}$
W	1	1	1	$\tau(b_4, \langle b_2, b_4\rangle)$	$\tau(b_4, \langle b_1, b_4\rangle)$	1

representation $\theta_{12}^{(3)}$ (from which the representation θ_1 was induced) and a representation $\theta_{12}^{(42)}$ supported by a 42-subspace $X_2^{(42)}$ of X which is the direct sum of 3-subspaces $X(U)$ taken for all 2-subspaces U of $V = V_5(2)$ which contain $V_1 = \langle b_5\rangle$ and differ from $V_2 = \langle b_4, b_5\rangle$. Clearly there are 14 such 2-subspaces, which gives the right dimension of $X_2^{(42)}$ (notice that $X_2^{(3)} := X(V_2)$ supports $\theta_{12}^{(3)}$). Our main objective is to determine the eigenspaces of R_2 in $X_2^{(42)}$. We show that the eigenspaces are one-dimensional and that their kernels are the hyperplanes from the orbit $\mathcal{P}_2$ as in Section 7.3.

We have

$$R_2 = \langle \tau_{4i}, \tau_{5i} \mid i = 1, 2, 3\rangle.$$

If we put $U^{(1)} = \langle b_5, b_3\rangle$ and $U^{(2)} = \langle b_5, b_3 + b_4\rangle$ then every transvection-generator of R_2, except for τ_{43}, stabilizes $U^{(1)}$ and $U^{(2)}$, while τ_{43} swaps these two subspaces. Therefore the 6-subspace

$$Y = X(U^{(1)}) \oplus X(U^{(2)})$$

is R_2-stable and we shall decompose Y into the direct sum of R_2-irreducibles. The subspace

$$W = \langle b_1, b_2, b_4\rangle$$

complements in $V = V_5(2)$ both $U^{(1)}$ and $U^{(2)}$, and W is stabilized by every transvection-generator of R_2. Therefore the action of an element $r \in R_2$ on Y can be recovered from its action on $U^{(1)}$, $U^{(2)}$ and W with application of the principles (i) to (iv) from the previous section. These actions are summarized in Table 7.1. In most cases the order of the action is given. The actions of τ_{41} and τ_{42} are described in terms of transvection. The last column indicates that τ_{43} swaps $U^{(1)}$ and $U^{(2)}$, centralizing W.

In view of the principles (i) to (iv) from the previous section it follows from Table 7.1 that both τ_{51} and τ_{52} centralize Y, while τ_{53} acts by negating every vector. The elements τ_{41} and τ_{42} generate on each $U^{(i)}$ an elementary abelian action of order 4.

Let $x_1^{(i)}$, $x_2^{(i)}$ and $x_3^{(i)}$ be eigenvectors of $\langle \tau_{41}, \tau_{42}\rangle$ in $U^{(i)}$ centralized by τ_{41}, τ_{42} and $\tau_{41}\tau_{42}$, respectively, where $i = 1$ or 2. Since τ_{43} is an involution

Table 7.2

	y_1^+	y_1^-	y_2^+	y_2^-	y_3^+	y_3^-
τ_{51}	1	1	1	1	1	1
τ_{52}	1	1	1	1	1	1
τ_{53}	−1	−1	−1	−1	−1	−1
τ_{41}	1	1	−1	−1	−1	−1
τ_{42}	−1	−1	1	1	−1	−1
τ_{43}	1	−1	1	−1	1	−1

commuting with $\langle \tau_{41}, \tau_{42} \rangle$, the eigenvectors $x_j^{(i)}$ can be chosen in such a way that τ_{43} swaps $x_j^{(1)}$ and $x_j^{(2)}$ for every $j = 1$, 2 and 3. Then the vectors

$$y_j^+ = x_j^{(1)} + x_j^{(2)} \text{ and } y_j^- = x_j^{(1)} - x_j^{(2)}$$

taken for $j = 1, 2, 3$ form a basis set of R_2-eigenvectors. The corresponding eigenvalues are as given in Table 7.2.

Let $P(y_1^+)$ be the hyperplane in R_2 which is the kernel at y_1^+. Then

$$P(y_1^+) = \langle \tau_{51}, \tau_{52}, \tau_{41}, \tau_{43}, \tau_{41}\tau_{51}, \tau_{42}\tau_{52}\tau_{43}\tau_{53} \rangle.$$

In terms of Sector 7.3 the first two generators are contained in $R_2^{(5)}$, the middle two in $R_2^{(4)}$ and the last two in $R_2^{(4\frac{1}{2})}$. Therefore $P(y_1^+)$ is contained in a G_2-orbit $\mathcal{P}_2$ of length 42. By the dimension consideration there is a G_2-bijection which maps $\mathcal{P}_2$ onto the kernels at the R_2-eigenvectors in $X_2^{(42)}$. Furthermore for the hyperplane $P(y_1^+)$ the lines M_j in the projective plane of $U_3 = \langle b_1, b_2, b_3 \rangle$ as defined at the end of Section 7.3 are

$$\langle b_1, b_2 \rangle, \quad \langle b_1, b_3 \rangle, \quad \langle b_1, b_2 + b_3 \rangle$$

for $j = 5$, 4 and $\frac{1}{2}$, respectively, thus comprising the triple of lines in the projective plane passing through b_1.

The next aim is to prove the following *stabilizer* property.

Lemma 7.5 *Let L^* be the normalizer of $L^{(5)}$ in G_{12}, so that $L^* \cong L_3(2) \times 2$. Then the centralizer of y_1^+ in G_{12} is the semidirect product of $P(y_1^+)$ and the centralizer of y_1^+ in L^*, isomorphic to D_8.*

Proof By Lemma 7.3 the stabilizer of $P(y_1^+)$ in G_2, modulo R_2, is isomorphic to S_4. The subgroup G_{12} has index 3 in G_2 and R_2 induces on the 1-subspace spanned by y_1^+ an action of order 2 with kernel $P(y_1^+)$. Thus, in order to justify the above, it is sufficient to produce elements in L^* which centralize y_1^+ and generate a subgroup D of order 8 (isomorphic to D_8). We take

$$D = \langle \tau_{12}\tau_{53}, \tau_{31}\tau_1, \tau_{32}\tau_2 \rangle,$$

where τ_1 and τ_2 are suitable elements from $\langle \tau_{51}, \tau_{52} \rangle$ which shift τ_{31} and τ_{32} inside the copy of $L^{(5)}$ which contains $\tau_{21}\tau_{53}$. The isomorphism $D \cong D_8$ is immediately evident from the commutator relations among the transvections. Also from these relations it follows that D normalizes $P(y_1^+)$ and it remains only to recover the action of the generators of D which centralize y_1^+. The involution τ_{12} centralizes each of $U^{(1)}$, $U^{(2)}$ and W, therefore it stabilizes Y. Furthermore, it centralizes τ_{41} and conjugates τ_{42} onto $\tau_{41}\tau_{42}$. Therefore, when acting on $X(U^{(i)})$, the involution τ_{12} stabilizes the eigenspace of τ_{41} (which is spanned by $x_i^{(1)}$) and swaps the eigenspace spanned by $x_i^{(2)}$ with that spanned by $x_i^{(3)}$. Since τ_{12} is an involution, by the S_4-*restriction* Lemma 7.1 τ_{12} negates $x_i^{(1)}$ for $i = 1$ and 2. Since τ_{53} negates the whole of Y, the product $\tau_{12}\tau_{53}$ centralizes $y_1^+ = x_1^{(1)} + x_1^{(2)}$ as claimed. The argument for τ_{31} and τ_{32} is straightforward. □

7.6 Lifting from G_{12} to G_2

In this section we define a 45-dimensional representation θ_2 of G_2, whose restriction to G_{12} coincides with the representation θ_{12} from the previous section. The desired representation θ_2 will be the direct sum of a three-dimensional representation $\theta_2^{(3)}$ and a 42-dimensional representation $\theta_2^{(2)}$, restricting to $\theta_{12}^{(3)}$ and $\theta_{12}^{(42)}$, respectively.

For the three-dimensional part the lifting is particularly easy, since we just enlarge the kernel from R_2 to the pre-image R_2^* of the normal subgroup of order 3 in $G_{12}/R_2 \cong L_3(2) \times S_3$. Since G_2/R_2^* is canonically isomorphic to G_{12}/R_2, the acting group $L_3(2) \times 2$ remains the same.

For the 42-dimensional part we construct $\theta_2^{(42)}$ as a representation induced from a non-trivial one-dimensional representation of the stabilizer $G_2(P)$ of a hyperplane $P \in \mathcal{P}_2$ of R_2. By the *hyperplane* property Lemma 7.3 the stabilizer $G_2(P)$ is the semidirect product of R_2 and an S_4-subgroup $S^{(c)}$ in the complement

$$M := N_{G_2}(L^{(5)}) \cong L_3(2) \times S_3.$$

Clearly $S^{(c)}$ is exactly the stabilizer of P in M. Therefore the quotient group $G_2(P)/P$ is a group $S_4 \times 2$, where the first direct factor is the isomorphic image of $S^{(c)}$ and the second direct factor is just the image of R_2/P. Let $S^{(d)}$ be the 'diagonal' S_4-subgroup in $G(P)/P$ (which is just the second complement to

R_2/P). Let φ_P be the one-dimensional representation of $G_2(P)$, whose kernel is the pre-image of $S^{(\mathrm{d})}$ (this pre-image clearly has index 2 in $G(P)$). Define $\theta_2^{(42)}$ to be the representation which is φ_P induced from $G_2(P)$ to G_2. We claim that $\theta_2^{(42)}$ when restricted to G_{12} coincides with $\theta_{12}^{(42)}$. In fact, the stabilizer of $P(y_1^+)$ in

$$N_{G_{12}}(L^{(5)}) \cong L_3(2) \times 2$$

isomorphic to D_8 intersects the centralizer of y_1^+ in an index 3-subgroup (which is elementary abelian). Therefore the D_8-subgroup is a Sylow 2-subgroup in the diagonal S_4-subgroup $S^{(\mathrm{d})}$ (subject to the identification of P with $P(y_1^+) = C_{R_2}(y_1^+)$). Therefore we have established the following lemma.

Lemma 7.6 *The representation $\theta_2^{(42)}$ of G_2 when restricted to G_{12} is equivalent to the representation $\theta_{12}^{(42)}$.* □

7.7 Hyperplanes in R_3 and Covers of G_3

Before expanding the representations θ_1 and θ_2 constructed in the previous section onto the whole of the Mathieu amalgam $\mathcal{A} = \{G_1, G_2, G_3\}$, we make explicit some properties of G_3 viewed as a completion of the amalgam

$$\mathcal{A}_3 = \{G_{13}, G_{23}\}.$$

The subgroups G_{13} and G_{23} can be defined as the normalizers of T_3 in G_1 and G_2, respectively. Here $T_3 = R_3 : \langle t \rangle$ is the core of the amalgam $\mathcal{A}_3$, where t is an element of order 3, which normalizes R_3 and acts on it fixed-point freely (we have taken $t = \tau_{12}\tau_{21}$). If we factorize the members of the amalgam $\mathcal{A}_3$ over the radical T_3, we obtain the simple amalgam

$$\mathcal{A}_3/T_3 = \{G_{13}/T_3, G_{23}/T_3\} \cong \{S_4 \times 2, S_4 \times 2\},$$

having $G_3/T_3 \cong S_6$ as a faithful generating completion. The coset graph of this completion is the generalized quadrangle Θ associated with $S_6 \cong Sp_4(2)$. The graph Θ is distance-regular with the following intersection diagram:

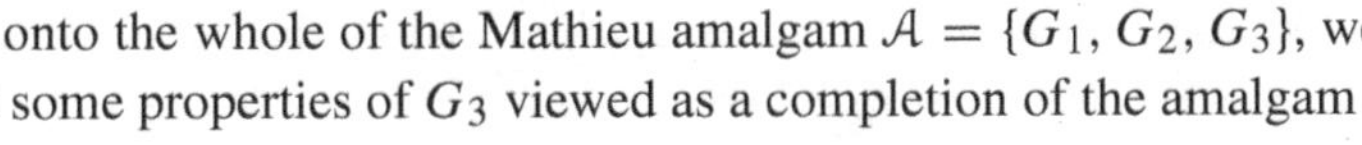

Let Ω be a set of size 6 on which G_3/T_3 acts as the symmetric group, such that G_{23}/T_3 stabilizes a 2-subset of Ω. One part of Θ (formed by the cosets of G_{23}) is the set of all 2-subsets of Ω, while the other part (the cosets of G_{13}) is the set of partitions of Ω into three disjoint 2-subsets. The adjacency relation between the parts is by inclusion. The parts of Ω are swapped by the outer automorphisms of $G_3/T_3 \cong S_6$, although these automorphisms do not extend to automorphisms of the whole Mathieu amalgam.

Since G_3 is the pre-image of the stabilizer of a hyperoval in

$$\mathrm{Aut}(T_3) \cong 2^6 : \Gamma_3(4),$$

we obtain the following *hyperoval–hyperplane* principle directly from the orbit description of the hyperplane stabilizer on the point-set of the projective plane of order 4.

Lemma 7.7 *The group G_3 acts on the set of hyperplanes in R_3 with two orbits $\mathcal{H}_1$ and $\mathcal{H}_3$, whose lengths are* 18 *and* 45. *The element t of order* 3 *extending R_3 to T_3 acts on the set of hyperplanes fixed-point freely. The set of orbits of t on $\mathcal{H}_2$ is naturally indexed by the 2-subsets of the six-element set Ω. The intersection of hyperplanes in the orbit stabilized by G_{13} is the subgroup $R_2 \cap R_3 = \langle \tau_{41}, \tau_{42}, \tau_{51}, \tau_{52} \rangle$.* □

To discuss the covers of G_3 which are completions of $\mathcal{A}_3$ with abelian kernels with respect to homomorphisms onto G_3, let us adopt the bar convention for the quotient over the radical T_3. Then

$$\bar{\mathcal{A}}_3 = \{\bar{G}_{13}, \bar{G}_{23}\} \cong \{S_4 \times 2, S_4 \times 2\}$$

is a simple amalgam of which $\bar{G}_3 \cong S_6$ is a faithful generating completion with coset graph Θ, which is the generalized quadrangle of order 2 with the above intersection diagram. Let $\bar{G}_3^{(\mathrm{u})}$ be the universal completion of $\bar{\mathcal{A}}_3$ and let

$$\eta^{(\mathrm{u})} : \bar{G}_3^{(\mathrm{u})} \to \bar{G}_3$$

be the homomorphism of the completion groups (which is the identity on $\bar{\mathcal{A}}_3$, viewed as a subamalgam in the completion groups). Then it follows from the standard theory of rank-2 amalgams[5] that the kernel $K^{(\mathrm{u})}$ of $\eta^{(\mathrm{u})}$ is freely generated by 16 fundamental cycles of Θ. The rank 16 of the free group $K^{(\mathrm{u})}$ comes from the number 45 of edges minus the number 30 of vertices plus 1. Let $K^{(\mathrm{a})}$ be the commutator subgroup of $K^{(\mathrm{u})}$, so that the completion group $\bar{G}_3^{(\mathrm{a})} = \bar{G}_3^{(\mathrm{u})}/K^{(\mathrm{a})}$ is the largest completion of $\bar{\mathcal{A}}_3$ subject to the condition that it possesses a homomorphism

$$\eta^{(\mathrm{a})} : \bar{G}_3^{(\mathrm{a})} \to \bar{G}_3,$$

[5] J.-P. Serre, *Arbres, amalgams, SL_2*, Astérisque 46, Société Mathématique de France, Paris, 1977.

whose kernel is an abelian group. Since Θ is the classical generalized quadrangle of $\bar{G}_3 \cong S_6 \cong Sp_4(2)$, the quotient $K^{(\mathrm{u})}/K^{(\mathrm{a})}$ is the 16-dimensional Steinberg module over the ring of integers. This module does not possess nontrivial homomorphisms into the permutation module of $\bar{G}_3$ on the vertex set of Θ and this is the reason for the following *no abelian kernel* property.

Lemma 7.8 *Let $\bar{G}_3^{(x)}$ be a completion group of $\bar{\mathcal{A}}_3$ which possesses a homomorphism onto $\bar{G}_3$ with an abelian kernel contained in the k-permutation module of $\bar{G}_3$ on $\binom{\Omega}{2}$. Then the kernel of the homomorphism is trivial and $\bar{G}_3^{(x)} = \bar{G}_3$.* □

The situation described in the above lemma is similar to those considered in Sections 6.12[6] and 2.11.[7] In the former case the situation occurred in the construction of the 1333-dimensional representation of the fourth Janko group, J_4, where the role of $\bar{G}_3$ was played by S_5 and Θ was the Petersen graph. In the latter case it was involved in the construction of the 196 883-dimensional representation of the Monster group, where $\bar{G}_3$ was the group $L_3(2)$ and Θ was the incidence graph of the projective plane of order 2.

In the current situation the argument runs as follows. The group $K^{(\mathrm{u})}$ (known also as the first homology group of Θ) is generated by the fundamental cycles. Let c be one of them, say

$$\begin{aligned} 12 &\to 12|34|56 \to 34 \to 34|16|25 \to 26 \\ &\to 35|16|24 \to 35 \to 12|35|46 \to 12. \end{aligned}$$

Then the permutations $g = (1\,4\,6\,5)(2\,3)$ and $h = (1\,4)(5\,6)(2\,3)$ of Ω stabilize c (as an undirected cycle), with g preserving its orientation and h reversing it. The crucial observation is that $\langle g\rangle$ and $\langle g, h\rangle$ have the same five orbits on $\binom{\Omega}{2}$. Let

$$\psi : K^{(\mathrm{u})}/K^{(\mathrm{a})} \to k^{\binom{\Omega}{2}}$$

be a $\bar{G}_3 \cong S_6$ homomorphism. Then $\psi(cK^{(\mathrm{a})})$ is g-invariant and, by the above observation, it is also h-invariant, which implies that $\psi(cK^{(\mathrm{a})}) = \psi(c^{-1}K^{(\mathrm{a})})$ and therefore ψ factorizes through $K^{(\mathrm{u})}/K^{(\mathrm{a})}$ taken modulo 2, although the latter is the irreducible 16-dimensional Steinberg module, which could not possibly be involved in $2^{\binom{\Omega}{2}}$ because of the dimension consideration.

6 A. A. Ivanov, J_4, Oxford University Press, Oxford, 2004.

7 A. A. Ivanov, *The Monster Group and Majorana Involutions*, Cambridge University Press, Cambridge, 2009.

Notice that $K^{(u)}/K^{(a)}$ taken modulo 3 is reducible. In particular, it contains a co-dimension-1 submodule responsible for the triple cover

$$3 \cdot S_6 \to S_6,$$

which is a homomorphism of $\bar{\mathcal{A}}_3$-completion groups.

7.8 Identifying G_3

By now we have constructed a pair of 45-dimensional k-representations θ_1 and θ_2 of G_1 and G_2, respectively, which coincide when restricted to the intersection G_{12}. Next we would like to extend the representation to the whole of the Mathieu amalgam $\mathcal{A} = \{G_1, G_2, G_3\}$ by following the familiar route. We show that the subgroup in $GL_{45}(k)$ generated by $\theta_1(G_{13})$ and $\theta_2(G_{23})$ is isomorphic to G_3 and realize this isomorphism as a representation θ_3 of G_3.

We start by decomposing the 45-dimensional space X supporting θ_1 and θ_2 into eigenspaces of R_3, which possesses the following generating set of transvections:

$$R_3 = \langle \tau_{51}, \tau_{52}, \tau_{41}; \tau_{42}, \tau_{31}, \tau_{32} \rangle.$$

Notice that R_3 stabilizes the 3-space $X(V_2)$ which supports $\theta_2^{(3)}$, with the former four generators acting trivially and the latter two inducing an elementary abelian group of order 4. Let $x_{V_2}^{(i)}$ be the eigenvectors of $\langle \tau_{31}, \tau_{32} \rangle$ with kernels τ_{31}, τ_{32} and $\tau_{31}\tau_{32}$ for $i = 1$, 2 and 3, respectively. Then the $x_{V_2}^{(i)}$s are also the eigenvectors of R_3 with clearly visible kernels. These three eigenvectors (when properly rescaled) are transitively permuted by the order-3 element $t = \langle \tau_{12}\tau_{21} \rangle$, which together with R_3 generates the group

$$T_3 = R_3 : \langle t \rangle,$$

which is the largest subgroup in G_{123} that is normal both in G_{13} and in G_{23}. Since the irreducible components of T_3 on X are pairwise non-isomorphic, by Schur's lemma $C_{GL(X)}(T_3)$ is abelian and by Lemma 7.8 we obtain the following lemma.

Lemma 7.9 *The subgroups $\theta_1(G_{13})$ and $\theta_2(G_{23})$ generate in $GL(X) \cong GL_{45}(k)$ the group isomorphic to G_3.* □

By Lemma 7.9 the restrictions of θ_1 to G_{13} and of θ_2 to G_{23} can be extended to a representation θ_3 of G_3. Then the representations θ_1, θ_2 and θ_3 of G_1, G_2 and G_3, respectively, coincide on the relevant intersections and we thus obtain the main result of the chapter.

Theorem 7.10 *The triple* $(\theta_1, \theta_2, \theta_3)$ *of representations determines a* 45-*dimensional representation of the Mathieu amalgam*

$$\mathcal{A} = \{G_1, G_2, G_3\}$$

and, therefore, a 45-*dimensional irreducible representation of the universal completion of* $\mathcal{A}$ *isomorphic to* M_{24}. □

8

The Held Group

By studying the minimal parabolic amalgam of the Mathieu group M_{24}, which we call the *tilde Mathieu amalgam*, we observe that the condition to be constrained possesses a second solution which leads to the *tilde Held amalgam.* The universal completion of the Held amalgam is the sporadic simple group discovered by D. Held.[1] In this way we obtain all three of the simple groups $L_5(2)$, M_{24} and He in which the centralizer of an involution is of the form $2_+^{1+6} : L_3(2)$ as universal completions of closely related amalgams.

8.1 The Tilde Mathieu Amalgam

The *tilde Mathieu amalgam*

$$\widetilde{\mathcal{A}} = \{G_{14}, G_{24}, G_3\}$$

was introduced and discussed in Section 5.7 and by Theorem 5.22. M_{24} is a faithful generated completion of this amalgam (which is in fact universal), while the corresponding coset geometry is the famous tilde geometry of M_{24} with the diagram

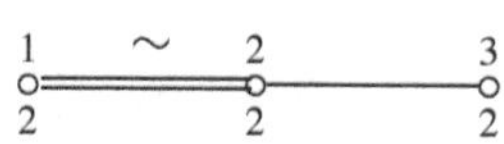

where the rightmost node is the coset geometry of the completion $3{\cdot}S_6$ of Goldschmidt's amalgam G_3^1, namely the remarkable triple cover of the classical generalized quadrangle of order 2 associated with

$$S_6 \cong Sp_4(2).$$

As in Section 4.1, let $V_5(2)$ be a five-dimensional $GF(2)$-space, let $B = (b_1, b_2, \ldots, b_5)$ be a basis of $V_5(2)$, let $(b_1^*, b_2^*, \ldots, b_5^*)$ be the basis of the dual

[1] D. Held, The simple groups related to M_{24}, *J. Algebra* **13** (1969), 253–296.

space of $V_5(2)$ such that $b_i^*(b_j) = \delta_{ij}$, and let $H = L_5(2)$ be the corresponding linear group. Let H_1 be the stabilizer of $V_1 = \langle b_5 \rangle$ in H (cf. Section 4.2). Let H_{14} be the stabilizer in H_1 of the hyperplane

$$V_4 = \langle b_2, b_3, b_4, b_5 \rangle$$

identified with the dual vector b_1^*. Then H_{14} is the centralizer in H of the transvection τ_{51}, which by definition centralizes V_4 and adds b_5 to every vector in $V_5(2) \setminus V_4$. The following result is straightforward.

Lemma 8.1 *It holds that*

$$H_{14} = C_H(\tau_{51}) \cong 2_+^{1+6} : L_3(2)$$

and the matrix realization of H_{14} in the basis B can be described as follows:

$$C_{L_5(2)}(\tau_{51}) = \begin{pmatrix} 1 & 0 & 0 & 0 & 0 \\ \tau_{21} & 1 & \tau_{23} & \tau_{24} & 0 \\ \tau_{31} & \tau_{32} & 1 & \tau_{34} & 0 \\ \tau_{41} & \tau_{42} & \tau_{43} & 1 & 0 \\ \tau_{51} & \tau_{52} & \tau_{53} & \tau_{54} & 1 \end{pmatrix}.$$

It is a well-known and remarkable fact that besides $L_5(2)$ there are exactly two further simple groups having the above group as the centralizer of an involution. These two groups are the sporadic simple groups Mathieu M_{24} and Held He.[2]

The radical $R_{14} = O_2(H_{14})$ is the product of the radical $R_1 = \langle \tau_{5i} \mid 1 \leq i \leq 5 \rangle$ of R_1 and the radical $R_4 = \langle \tau_{j1} \mid 1 \leq j \leq 5 \rangle$ of the stabilizer $H_4 \cong 2^4 : L_4(2)$ of V_4. Since

$$[\tau_{5i}, \tau_{j1}] = \tau_{51}^{\delta_{ij}},$$

R_{14} is indeed extraspecial of order 2^7 of plus type, while

$$K_{14} = \langle \tau_{32}, \tau_{42}, \tau_{43}, \tau_{23}, \tau_{34} \rangle$$

is the Levi complement to R_{14} in H_{14}. It is clear from the matrix realization that K_{14} is the stabilizer of the flag (b_1^*, b_5) together with the flag (b_5^*, b_1) opposite to it, so that

$$K_{14} = C_H(\tau_{51}) \cap C_H(\tau_{15}).$$

If we put $H_{24} = H_2 \cap H_4$, then the amalgam $\{H_{14}, H_{24}\}$ is not simple: it is contained in H_4 and contains R_4, so that the latter radical is normal in both H_{14} and H_{24}.

[2] D. Held, The simple groups related to M_{24}, *J. Algebra* **13** (1969), 253–296.

On the other hand, we can apply the automorphism α of $H'_{12} \cong 2^6 : L_3(2)$ which induces the permutation

$$\widehat{\alpha} = (L\, L^{(5)})(L^{(4)}\, L^{(4\frac{1}{2})})$$

to obtain a simple amalgam $\widetilde{\mathcal{X}} = \{G_{14}, G_{24}\}$, where G_{14} is identified with H_{14} and $G_{24} = H_{24}^{\alpha}$. The specific version of α which we have chosen commutes with

$$R_2 = \langle \tau_{4i}, \tau_{5j} \mid 1 \le i, j \le 3 \rangle$$

and shifts the generators of the complement L onto that of $L^{(5)}$ in the following manner:

$$\alpha : \tau_{21} \mapsto \tau_{21}\tau_{53}, \quad \alpha : \tau_{31} \mapsto \tau_{31}\tau_{52},$$
$$\alpha : \tau_{32} \mapsto \tau_{32}\tau_{51}\tau_{52}, \quad \alpha : \tau_{23} \mapsto \tau_{23}\tau_{51}\tau_{53}.$$

It is useful to keep in mind the following arrangement of the generators of G_2, which was introduced in Section 4.6:

$$G_2 = \begin{pmatrix} 1 & \tau_{12}\tau_{53} & \tau_{13}\tau_{52}\tau_{53} & 0 & 0 \\ \tau_{21}\tau_{53} & 1 & \tau_{23}\tau_{51}\tau_{53} & 0 & 0 \\ \tau_{31}\tau_{52} & \tau_{32}\tau_{51}\tau_{52} & 1 & 0 & 0 \\ \tau_{41} & \tau_{42} & \tau_{43} & 1 & \tau_{45}^{\alpha} \\ \tau_{51} & \tau_{52} & \tau_{53} & \tau_{54} & 1 \end{pmatrix}.$$

In the above matrix the top left 3×3 block contains the generators of $L^{(5)}$.

Viewing $\widetilde{\mathcal{X}} = \{G_{14}, G_{24}\}$ as a subamalgam in the Mathieu amalgam $\mathcal{A} = \{G_1, G_2, G_3\}$, we enlarge $\mathcal{X}$ by adjoining G_3, which is generated by $\mathcal{A}$:

$$\widetilde{\mathcal{A}} = \{G_{14}, G_{24}, G_3\}.$$

Note that G_3 is generated by its intersections with G_{14} and G_{24}, which are the normalizers of R_3 in the respective members of $\widetilde{\mathcal{X}}$.

8.2 Dichotomy

In Lemma 4.11 the Mathieu amalgam $\mathcal{A} = \{G_1, G_2, G_3\}$ was redefined as follows:

$$\mathcal{A} = \{G_1, G_2, X_3/C_{X_3}(T_3)\},$$

where X_3 is the universal completion of the amalgam $\mathcal{X}_3 = \{G_{13}, G_{23}\}$ of the intersections having $T_3 \cong 2^6 : 3$ as the radical. This redefinition was possible because the radical is non-abelian and contains its centralizer in the completion G_3 in question. In the case of the tilde Mathieu amalgam the radical of the amalgam $\widetilde{\mathcal{X}}_3 = \{G_{134}, G_{234}\}$ is R_3 which is abelian, although it is still self-centralized in G_3. Therefore $\widetilde{\mathcal{A}}$ can be redefined as

$$\widetilde{\mathcal{A}} = \{G_{14}, G_{24}, \widetilde{X}_3/E\},$$

where $\widetilde{X}_3$ is the universal completion of the amalgam $\widetilde{\mathcal{X}}_3$ and E is a complement to R_3 in $C_{\widetilde{X}_3}(R_3)$. Motivated by Theorem 2.2, we intend in this chapter to prove the following *dichotomy* principle.

Theorem 8.2 *The universal completion $\widetilde{X}_3$ of the amalgam $\mathcal{X}_3 = \{G_{143}, G_{234}\}$ contains precisely two complements $E^{(1)}$ and $E^{(2)}$ to R_3 in $C_{\widetilde{X}_3}(R_3)$. Define*

$$\mathcal{D}^{(i)} = \{G_{134}, G_{234}, \widetilde{X}_3/E^{(i)}\}.$$

Then, subject to renumbering, $\mathcal{D}^{(1)}$ is the tilde Mathieu amalgam $\widetilde{\mathcal{A}}$, whose universal completion is the Mathieu group M_{24}, while the universal completion of $\mathcal{D}^{(2)}$ is the Held sporadic simple group He.

Although the amalgams $\mathcal{D}^{(1)}$ and $\mathcal{D}^{(2)}$ are non-isomorphic,

$$\widetilde{X}_3/E^{(1)} \cong \widetilde{X}_3/E^{(2)} \cong G_3 \cong 2^6 : 3 \cdot S_6.$$

Since the tilde Mathieu amalgam $\widetilde{\mathcal{A}}$ exists, there is at least one complement $E^{(1)}$ to R_3 in $C_{\widetilde{X}_3}(R_3)$. Suppose that there is another complement, say $E^{(2)}$. Then, for $F := \widetilde{X}_3/(E^{(1)} \cap E^{(2)})$, we have

$$F \cong (2^6 \times 2^6) : 3 \cdot S_6,$$

with $O_2(F)$ containing three 2^6-subgroups that are normal in F, which are the images of R_3, $E^{(1)}$ and $E^{(2)}$. Therefore

$$\widetilde{X}_3/(E^{(1)} \cap E^{(2)})R_3 \cong \widetilde{X}_3/E^{(1)} \cong \widetilde{X}_3/E^{(2)} \cong G_3.$$

Furthermore, $G_3/R_3 \cong 3 \cdot S_6$ is a faithful generated completion of the amalgam

$$\widetilde{\mathcal{X}}_3/R_3 := \{G_{134}/R_3, G_{243}/R_3\} \cong \{S_4 \times 2, S_4 \times 2\}$$

(which is Goldschmidt's amalgam G_3^1).

8.3 G_3^1-Subamalgams in G_3

By the previous section, in order to prove the dichotomy principle Theorem 8.2 we have to classify the completions of Goldschmidt's amalgam $G_3^1 \cong \{S_4 \times 2, S_4 \times 2\}$ isomorphic to $G_3 \cong 2^6 : 3 \cdot S_6$. This is certainly equivalent to the classification (up to conjugation) of the subamalgams $\mathcal{T} = \{T_1, T_2\}$ in G_3 followed by selection from them of the generating ones.

On the other hand, it is easy to produce a non-generating subamalgam. In fact, let t be an element of order 3 in T_3 (our choice is $t = \tau_{12}\tau_{21}$). Then, since t acts fixed-point freely on R_3, and since $\langle t \rangle$ is a Sylow 3-subgroup of T_3, by the Frattini argument

$$\mathcal{T}^{(0)} := \{T_1^{(0)} = G_{134} \cap N_{G_3}(\langle t \rangle),\ T_2^{(0)} = G_{234} \cap N_{G_3}(\langle t \rangle)\}$$

is isomorphic to $\widetilde{\mathcal{X}}_3/R_3$, and therefore to G_3^1. We shall obtain an explicit description of $\mathcal{T}^{(0)}$ in order to analyze its possible "shifting" to a generating subamalgam of the same isomorphism type.

Recall that t induces on R_3 the structure of a three-dimensional $GF(4)$-space, the corresponding projective plane Ξ of order 4 has the orbits of t on the set of non-identity elements of R_3 as points, and

$$\mathrm{Aut}(\Xi) \cong P\Gamma L_3(4),$$

while $G_3/R_3 \cong 3 \cdot S_6$ is the stabilizer in that automorphism group of a hyperoval Θ.

Looking at the matrix expression for $G_{12} = H_{12} = C_{L_5(2)}(\tau_{51})$ we observe that

$$T_1^{(0)} = \langle \tau_{21}, \tau_{53}, \tau_{54}, \tau_{43}, \tau_{34} \rangle.$$

Furthermore, $Z_1^{(0)} = \langle \tau_{21} \rangle$ is the centre of $T_1^{(0)}$, $K_1^{(0)} = \langle \tau_{53}, \tau_{54} \rangle$ is the Klein 4-subgroup and $S_1^{(0)} = \langle \tau_{43}, \tau_{34} \rangle$ is an S_3-complement in the S_4-direct factor generated by all the generators except for τ_{21}. Considering the action of $T_1^{(0)}$, we observe that

$$C_1^{(0)} := C_{R_3}(Z_1^{(0)}) = \langle \tau_{31}, \tau_{41}, \tau_{51} \rangle$$

is elementary abelian of order 8 (since τ_{21} is a field automorphism in $\mathrm{Aut}(R_3T_3) \cong \Gamma L_3(4)$ and $T_1^{(0)}$ acts on $C_1^{(0)}$ as the stabilizer of the point τ_{51} in the corresponding Fano plane (with kernel $Z_1^{(0)}$ and image isomorphic to S_4)).

Similarly,

$$T_2^{(0)} = \langle \tau_{21}\tau_{53}, \tau_{43}, \tau_{53}, \tau_{54}, \tau_{45}^{\alpha} \rangle,$$

where α is the chosen isomorphism of H_{12} onto G_{12}. Notice that α commutes with R_2, and maps the Levi complement L onto the complement $L^{(5)}$ such that $R_1L = R_1L^{(5)}$, where R_1 is viewed as $C_{H_{12}}(\tau_{54})$. Furthermore, α sends $\tau_{21} \in L$ onto $\tau_{21}\tau_{53} \in L^{(5)}$ and

$$C_{G_{12}}(L^{(5)}) = \langle \tau_{54}, \tau_{45}^{\alpha} \rangle \cong S_3.$$

Then, adopting the notation introduced for $T_1^{(0)}$, we have $Z_2^{(0)} = \langle \tau_{21}\tau_{53} \rangle$, $K_2^{(0)} = \langle \tau_{43}, \tau_{53} \rangle$, $S_2^{(0)} = \langle \tau_{54}, \tau_{45}^{\alpha} \rangle$ and $C_2^{(0)} = \langle \tau_{41}, \tau_{51}, \tau_{31}\tau_{52} \rangle$. The group $T_2^{(0)}$ acts on $C_2^{(0)}$ with kernel $Z_2^{(0)}$ inducing the stabilizer of the line $\{\tau_{41}, \tau_{51}, \tau_{41}\tau_{51}\}$ in the Fano plane of $C_2^{(0)}$.

We are looking for subamalgams

$$\mathcal{T}^{(s)} = \{T_1^{(s)}, T_2^{(s)}\}$$

in G_3, isomorphic to $\mathcal{T}^{(0)}$ and having the same image in $G_3/R_3 \cong 3 \cdot S_6$. Thus there should be an isomorphism

$$\sigma : \mathcal{T}^{(0)} \to \mathcal{T}^{(s)}$$

such that τ and $\sigma(\tau)$ induce the same automorphism of R_3, meaning that

$$\sigma(\tau) = \tau s(\tau)$$

for some $s(\tau) \in C_{G_3}(R_3) = R_3$. Clearly

$$s : \mathcal{T}^{(0)} \to R_3,$$

defined via $\tau \mapsto s(\tau)$, is a cocycle, when restricted to each of $T_1^{(0)}$ and $T_2^{(0)}$, which means that the equality

$$s(t_1 t_2) = s(t_1)^{t_2} s(t_2)$$

holds for $t_1, t_2 \in T_i^{(0)}$ when $i = 1$ and 2.

8.4 Constructing the Cocycle

In this section we construct a cocycle $c: \mathcal{T}^{(0)} \to R_3$ subject to the restriction that for every $\tau \in T_i^{(0)}$ the inclusion

$$c(\tau) \in C_i^{(0)}$$

holds.

Subject to this restriction we are able to find a unique non-zero cocycle, although it is not the one we are after, since it leads to an amalgam $\mathcal{T}^{(c)}$ conjugate to $\mathcal{T}^{(0)}$ by the element

$$\tau_{31}\tau_{41}\tau_{51}\tau_{52}.$$

We find it easier to start with $i = 2$. In this case (unlike the $i = 1$ case) the action of $T_2^{(0)}$ on $C_2^{(0)}$ is fixed point free, so that

$$Z(T_2^{(0)} C_2^{(0)}) = Z_2^{(0)},$$

which forces the equality

$$c(\tau_{21}\tau_{53}) = 1.$$

Next, since both $r = \tau_{45}^{\alpha}\tau_{54}$ and $r^{(c)} := \tau_{45}^{\alpha}c(\tau_{45}^{\alpha})\tau_{54}c(\tau_{54})$ are order 3 elements and each of them generates a Sylow 3-subgroup in

$$\langle C_2^{(0)}, r\rangle = \langle C_2^{(0)}, r^{(c)}\rangle \cong 2^3 : 3$$

we can always conjugate $r^{(c)}$ onto r by an element from $C_2^{(0)}$ and because of this we assume without loss of generality that $r^{(c)} = r$.
The equality $r = r^{(c)}$ leaves the possibility that $c(\tau_{45}^{\alpha})$ and $c(\tau_{54})$ project properly onto

$$C_{C_2^{(0)}}(r) = \langle \tau_{31}\tau_{52}\rangle.$$

But this possibility should be ruled out, since $\tau_{45} \in T_1^{(0)}$ while $\tau_{31}\tau_{52} \notin C_1^{(0)}$. This forces $c(\tau_{54})$ to be the remaining element of $C_2^{(0)}$ it commutes with:

$$c(\tau_{45}) = \tau_{51}.$$

Since τ_{54} and τ_{45}^{α} generate S_3, which acts naturally on $\langle \tau_{41}, \tau_{51}\rangle$, we must have $\tau_{45}^{\alpha} = \tau_{41}$.
Now it remains to determine the values of the cocycle c on the elements of the Klein four subgroup $K_2^{(0)}$. To achieve this we apply the fact that $K_2^{(c)} \leq K_2^{(0)}[r, C_2^{(0)}]$ and the commutator relations

$$1 = [\tau_{45}^{\alpha}, \tau_{43}] = [\tau_{45}^{\alpha}c(\tau_{45}^{\alpha}), \tau_{43}c(\tau_{43})],$$
$$1 = [\tau_{54}, \tau_{53}] = [\tau_{54}c(\tau_{54}), \tau_{53}c(\tau_{53})].$$

This gives

$$c(\tau_{43}) = \tau_{41}, \quad c(\tau_{53}) = \tau_{51}.$$

At this point the construction of the cocycle on $T_1^{(0)}$ is complete. We know the cocycle on $T_1^{(0)} \cap T_2^{(0)}$, so that $T_1^{(c)} \cap T_2^{(c)}$ is the direct product of the $\langle \tau_{21}\tau_{53}\rangle$ and the dihedral group of order 8 with centre generated by $\langle \tau_{53}\tau_{41}\rangle$. Since $Z(T_1^{(c)} \cap T_2^{(c)})$ contains the group $Z_2^{(c)}$ generated by $\tau_{21}c(\tau_{21})$ we conclude that

$$c(\tau_{21}) = \tau_{51}$$

and it only remains to determine $c(\tau_{34})$ and it is easy to see that it must be τ_{31}. Thus we have proved that cocycle c exists and that it is unique subject to the non-triviality assumption.

Lemma 8.3 *The cocycle c subject to the constrain imposed at the beginning of the section, is uniquely determined:*

$$c : \tau_{21} \mapsto \tau_{51}, \quad c : \tau_{34} \mapsto \tau_{31}, \quad c : \tau_{43} \mapsto \tau_{41},$$
$$c : \tau_{45}^{\alpha} \mapsto \tau_{41}, \quad c : \tau_{53} \mapsto \tau_{51}, \quad c : \tau_{54} \mapsto \tau_{51}. \qquad \square$$

As we have pointed out, the cocycle in Lemma 8.3 is not the generating one. But the generating one must exist due to the existence of two rank 3 tilde geometries with isomorphic subamalgams formed by the first two members. On the other hand, D. Holt,[3] using his cohomology computer package, has proved the following.

Lemma 8.4 *The group $G_3 \cong 2^6 : 3 \cdot S_6$ contains a unique class of generating subalamgams $\mathcal{T}^{(s)}$. Therefore the* generating *cocycle exists and is unique up to conjugation.* $\square$

The above lemma also follows from direct application of Goldschmidt's lemma to an amalgam of the same type with $\{G_{134}, G_{234}\}$ kindly performed by Sergey Shpectorov[4].

8.5 Two Completions

By the existence and uniqueness of the generating cocycle s implied by Lemma 8.4, we know that R_3 possesses exactly two complements, $E^{(1)}$ and $E^{(2)}$, to R_3 in the centralizer of R_3 in the universal completion of $\widetilde{\mathcal{X}}_3 = \{G_{134}, G_{234}\}$. Furthermore, we can produce an explicit construction of the quotient F of the universal completion over the intersection of the two complements.

Take two isomorphic copies of G_3:

$$G_3^{(1)} = \{g^{(1)} \mid g \in G_3\} \text{ and } G_3^{(2)} = \{g^{(2)} \mid g \in G_3\},$$

where $g \mapsto g^{(1)}$ and $g \mapsto g^{(2)}$ are isomorphisms of G_3 onto $G_3^{(1)}$ and $G_3^{(2)}$, respectively. Let D be the direct product of these two groups:

$$D = \{(g^{(1)}, h^{(2)}) \mid g, h \in G_3\}.$$

If t runs through the elements of the amalgam $\mathcal{T}^{(0)}$ and r runs through R_3, it is easy to see that the elements

$$a(t, r) := ((tr)^{(1)}, (ts(t)r)^{(2)})$$

[3] D. Holt, private communication, March 2018.
[4] S. Shpectorov, private communication, March 2018.

run though a subamalgam $\widetilde{\mathcal{Y}}_3$ in $G_3^{(1)} \times G_3^{(2)}$ that is isomorphic to $\widetilde{\mathcal{X}}_3$ with

$\{a(t,r) \mid t \in T_1^{(0)}, r \in R_3\} \cong G_{134}$ and $\{a(t,r) \mid t \in T_2^{(0)}, r \in R_3\} \cong G_{234}$,

with $R^{(3)} := \{(r^{1)}, r^{(2)}) \mid r \in R_3\}$ being the radical of $\widetilde{\mathcal{Y}}_3$. The subamalgam $\widetilde{\mathcal{Y}}_3$ generates in D a subgroup $F \cong (2^6 \times 2^6) : 3 \cdot S_6$. This subgroup is a subdirect product in D with respect to a pair of homomorphisms of $G^{(1)}$ and $G_2^{(2)}$ onto $G_3/R_3 \cong 3 \cdot S_6$. This follows from the fact that $\mathcal{T}^{(s)}$ generates the whole of G_3, and it is so since the elements $a(t,r)$ are a sort of diagonal with respect to the isomorphism of $\mathcal{T}^{(0)}$ onto $\mathcal{T}^{(s)}$ performed by the mapping

$$t \mapsto ts(t).$$

Now, identifying $\widetilde{\mathcal{Y}}_3$ with the subamalgam $\{G_{134}, G_{234}\}$ in the amalgam $\{G_{14}, G_{24}\}$, we obtain two amalgams

$$\widetilde{\mathcal{A}} = \{G_{14}, G_{24}, F/R_3^{(2)}\} \text{ and } \widetilde{\mathcal{H}} = \{G_{14}, G_{24}, F/R_3^{(1)}\}.$$

The first amalgam is indeed isomorphic to the tilde Mathieu amalgam. In fact, the subamalgam

$$\{(t^{(1)}, (ts(t))^{(2)}) \mid t \in \widetilde{\mathcal{T}}^{(0)}\}$$

modulo $R_3^{(2)}$ maps onto a copy of $\mathcal{T}^{(0)}$ which normalizes a subgroup of order 3 in $R_3 T_3$, and this happens in the Mathieu amalgam (whence $\mathcal{T}^{(0)}$ was taken in the first place), while modulo $R_3^{(1)}$ it maps onto the shifted subamalgam $\mathcal{T}^{(s)}$, which generates the whole of G_3. Therefore $\widetilde{\mathcal{A}}$ and $\widetilde{\mathcal{H}}$ have isomorphic subamalgams formed by their first two members, and their third members are isomorphic to $G_3 \cong 2^6 : 3 \cdot S_6$, although the amalgams are not isomorphic. Since $\widetilde{\mathcal{A}}$ is contained in the Mathieu amalgam $\mathcal{A}$, it possesses M_{24} as a faithful generating completion. This completion is universal, which is not at all obvious. For the amalgam $\widetilde{\mathcal{H}}$ even the existence of a faithful completion is problematic at this stage. To identify the universal completions of $\widetilde{\mathcal{A}}$ and $\widetilde{\mathcal{H}}$ with M_{24} and the Held sporadic simple group He, respectively (and thus to complete the proof of the dichotomy), we need to apply the technique of generators and relations. In order to accomplish this task we start in the next section with a presentation of the triple cover $3 \cdot S_6$ as a completion of the G_3^1-amalgam $\mathcal{T}^{(0)}$.

8.6 Presenting $3 \cdot S_6$

Let $T^{(0)} = N_{G_3}(\langle t \rangle) \cong 3 \cdot S_6$, where t is a generator of a Sylow 3-subgroup in $R_3 T_3 \cong 2^6 : 3$, where as before we take $t = \tau_{12}\tau_{21}$. Then $T^{(0)}$ is a faithful generating completion of the amalgam

$$\mathcal{T}^{(0)} = \{T_1^{(0)}, T_2^{(0)}\} \cong \{S_4 \times 2, S_4 \times 2\}.$$

In Section 8.3 we chose generators for the members of $\mathcal{T}^{(0)}$ and, by uniting them, we now obtain the following set of generators of $T^{(0)}$:

$$T^{(0)} = \langle \tau_{21}, \tau_{53}, \tau_{54}, \tau_{43}, \tau_{34}, \tau_{45}^{\alpha} \rangle.$$

The first four generators are for the intersection $T_1^{(0)} \cap T_2^{(0)} \cong 2 \times D_8$. By adjoining the fifth or sixth generator to the intersection we obtain $T_1^{(0)}$ or $T_2^{(0)}$, respectively. It follows immediately that we can write down a presentation for the amalgam making use either of the commutator and product order relations for the generators which are transvections or of the description of the subgroup structure of the $T_i^{(0)}$s in Section 8.3. Thus it remains only to produce enough relations involving each of the last two generators, which are τ_{34} and τ_{45}^{α}. This amounts to the following *three-pentagon* principle.

Lemma 8.5 *With respect to the above six-element generating set, the group $T^{(0)}$ is given by a presentation of the amalgam $\mathcal{T}^{(0)}$ together with the relations*

$$(\tau_{34}\tau_{45}^{\alpha})^5 = 1, \quad (\tau_{34}\tau_{43}\tau_{45}^{\alpha})^5 = 1, \quad (\tau_{54}\tau_{45}^{\alpha}\tau_{43}\tau_{34})^5 = 1.$$

Proof A possible way to check this statement is to run a coset enumeration program of a computer available in every group-theoretical package. In order to get a conceptual computer-free justification one might proceed along the following lines. First we need to check that these relations hold in $T^{(0)}$ and then justify that, together with the presentation of the amalgam, they provide a presentation for that group.

To check the relations in $T^{(0)}$, we calculate the orders of the automorphisms of R_3 induced by the products of the generators in the brackets, and in all three cases the order turns out to be 5. Since $T^{(0)}$ normalizes $T_3 = \langle t \rangle$, whose action on R_3 is fixed-point free, only the identity element in $T^{(0)}$ centralizes R_3, so that the relations do indeed hold in $T^{(0)}$. The action of the generators on R_3 can be easily deduced from the presentation of the amalgam $\mathcal{X} = \{G_1, G_2\}$, and we will reveal the details of this calculation in the next section when we produce a presentation for $\widetilde{\mathcal{H}}$ while calculating the behaviour of the corresponding images of the cocycle s under the products.

To justify the sufficiency of these relations, we take a look at the coset graph of $T^{(0)}$ as a completion of $\widetilde{\mathcal{T}}^{(0)}$. The coset graph is the Foster graph on 90 vertices, which is the triple cover of the generalized quadrangle of order 2. The distance-2 graph consists of two connected components. This components are isomorphic, and each of them is distance-transitive with the intersection array.

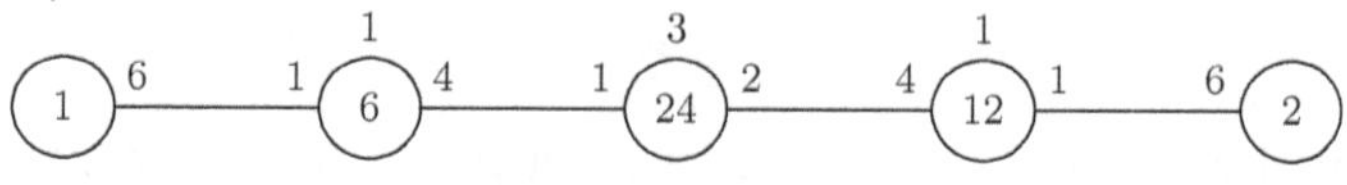

The isomorphism of the components is performed by an outer automorphism of S_6 lifted to an automorphism of $3 \cdot S_6$. This automorphism permutes $T_1^{(0)}$ and $T_2^{(0)}$, and therefore it does not extend to an automorphism of $\mathcal{X}$, although, due to its existence, the choice of the part is not relevant.

By the simple connectedness principle we have to adjoin enough relations to guarantee the existence in the coset graph of cycles whose homotopy classes generate the fundamental group of the graph. From the above diagram, we clearly see that every edge of the graph is contained in a unique triangle and any two vertices at distance 2 are contained in precisely three cycles of length 5. The triangles are already guaranteed by the amalgam and the relations in the three-pentagon principle, Lemma 8.5, impose the three classes of 5-cycles. Notice that, if we put

$$a_2 = \tau_{34}\tau_{43}, \; b_2 = \tau_{45}^{\alpha}, \; a_3 = \tau_{54}\tau_{45}^{\alpha}\tau_{43}, \; b_3 = \tau_{34},$$

then, by the relations in the amalgam, the order of a_2 and a_3 is 3, while the order of b_2 and b_3 is 2. Therefore, the last two relations in the three-pentagon principle, Lemma 8.5, together with relations implied by the amalgam provide the following famous presentation for A_5:

$$A_5 \cong \langle a_i, b_i \mid a_i^3 = b_i^2 = (a_i b_i)^5 = 1 \rangle.$$

Therefore, the last two relations ensure the existence of a pair of A_5-subgroups in the completion of $\mathcal{T}^{(0)}$ where these relations hold. These two A_5-subgroups represent different conjugacy classes of $3 \cdot S_6$ The subgraphs induced by the orbits of these two subgroups together with the 5-cycle imposed by the first relations cover representatives of all of the classes of 5-cycles in the considered bipartite part of the coset graph of $T^{(0)} \cong 3 \cdot S_6$.

The final step is to justify the claim that the 3- and 5-cycles in the bipartite part of the triple cover of the generalized quadrangle of order 2 generate the fundamental group of the graph. This fact can be proved using the standard procedure of reducing the cycles by making use of the fact that the subgraph induced by the vertices at distance 2 from a given one is connected. □

8.7 Presentations for M_{24} and He

The information we have deduced enables us to write down presentations for the amalgams $\widetilde{\mathcal{A}}$ and $\widetilde{\mathcal{H}}$. In both cases we start with a presentation of the amalgam $\widetilde{\mathcal{X}}_3 = \{G_{134}, G_{234}\}$. The generators are the relevant transvections τ_{ij} together with the twisted transvection τ_{45}^{α}. The relations are the usual commutator and product order relations among the transvections, together with the conjugation action of τ_{45}^{α} on the generating transvections of G_{24} visible from the matrix arrangement. In order to get the third member, we have to relate the generators $\tau_{45}^{\alpha} \in G_{234} \setminus G_{1234}$ and $\tau_{34} \in G_{134} \setminus G_{1234}$ so that they define the action of $3 \cdot S_6$ on R_3. By the three-pentagon principle, Lemma 8.5, this can be achieved by equalizing the left-hand sides of the three relations to suitable elements in R_3.

In the case of the M_{24}-tilde amalgam $\widetilde{\mathcal{A}}$ the subgroup R_3 is 'complemented' by the subamalgam $\mathcal{T}^{(0)}$, which normalizes a subgroup $\langle t \rangle$ of order 3, while $\langle t \rangle$ acts on R_3 fixed-point freely. Therefore in this case the equalities are 'pure', in the sense that on the right-hand side we always have the identity element. We summarize this as follows.

Theorem 8.6 *A presentation of the Mathieu tilde amalgam $\widetilde{\mathcal{A}}$ is obtained by adjoining the following relations to the presentation of $\{G_{134}, G_{234}\}$:*

$$(\tau_{34}\tau_{45}^{\alpha})^5 = 1, \quad (\tau_{34}\tau_{43}\tau_{45}^{\alpha})^5 = 1, \quad (\tau_{54}\tau_{45}^{\alpha}\tau_{43}\tau_{34})^5 = 1.$$

This is a presentation for the Mathieu group M_{24}. □

The intersection diagram of the point graph of the tilde geometry is given in Figure 8.1. It would be very desirable to deduce the simple connectedness of

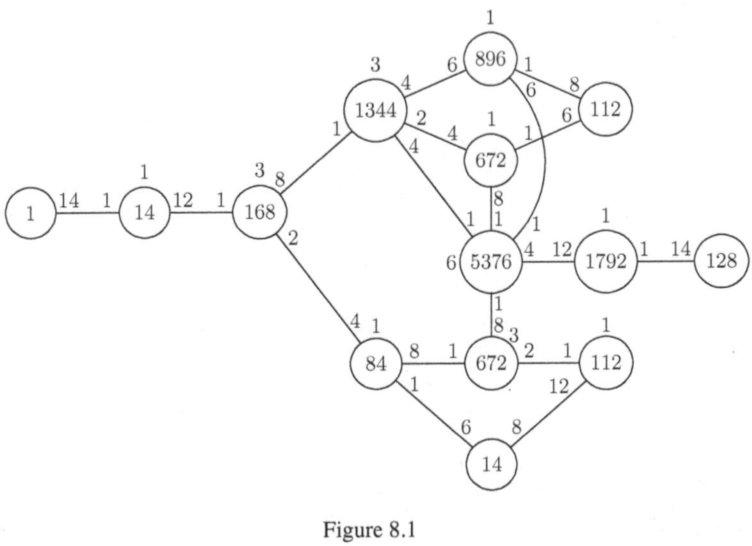

Figure 8.1

the tilde geometry of M_{24} by analysing cycles in that graph. A similar question applies to the tilde geometry of the Held group, but probably it is even more complicated.

In the case of the Held amalgam $\widetilde{\mathcal{H}}$ the subamalgam $\mathcal{T}^{(s)}$ is 'shifted' by the cocycle s, so, in order to obtain the relation, in each of the three relations in the three-pentagon principle, Lemma 8.5, one should substitute every generator t by $ts(t)$, evaluate the corresponding element of R_3 and place it on the right-hand side of the relation.

A presentation for $\widetilde{\mathcal{H}}$ can be found in[5] where the twisting automorphism is chosen to be different.

[5] M. Giudici, A.A. Ivanov, L. Morgan and C.E. Praeger, A characterisation of weakly locally projective amalgams related to A_{16} and the sporadic simple groups M_{24} and He. *J. Algebra*, **460** (2016), 340–365.

9

The Inevitability of Mathieu Groups

Primarily, the Mathieu groups are inevitable in the theory and classification of finite simple groups. Richard Brauer, in his pioneering work on the classification announced at the International Congress of Mathematicians in Amsterdam in 1954, had proved that whenever the centralizer of an involution in a finite simple group G is isomorphic to $GL_2(q)$ then either $G \cong L_3(q)$ or $q = 3$ and $G \cong M_{11}$ and 'foreshadowed the fascinating fact that conclusions of general classification theorems would necessarily include sporadic simple groups as exceptional cases'.[1] In the first section of this chapter we show how M_{11} appeared in another sporadic simple group that was discovered by O'Nan, while two further sections are devoted to a totally unexpected appearance of the Mathieu groups in areas very distant from finite group theory.

9.1 M_{11} in O'Nan's Group

By the middle of the 1980s the problem of whether a sporadic simple group H is a section (that is, a factor group of a subgroup) in a sporadic simple group G had not been resolved for all the pairs of 26 sporadic simple groups.[2] One of the outstanding cases was the involvement of the smallest Mathieu group $H = M_{11}$ in the group $G = ON$ discovered by Michael O'Nan.[3] The affirmative answer to the question was given through the construction of a really remarkable geometry $\mathcal{G}(ON)$ associated with O'Nan's group.[4] The geometry $\mathcal{G}(ON)$ is described by the diagram

1 D. Gorenstein, *Finite Simple Groups. An Introduction to Their Classification*, Springer, Berlin, 1982.

2 R. L. Griess, The friendly giant, *Invent. Math.* **69** (1982), 1–102.

3 M. E. O'Nan, Some evidence for the existence of a new simple group, *Proc. London Math. Soc.* **32** (1976), 421–479.

4 A. A. Ivanov and S. V. Shpectorov, A geometry for the O'Nan–Sims group connected to the Petersen graph, *Russian Math. Surveys* **41** (1986), 211–212.

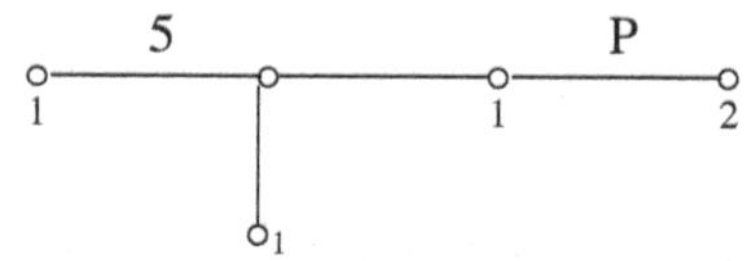

where the rightmost edge is the geometry of vertices and edges of the Petersen graph, the leftmost edge is the geometry of (the vertices and edges of) the pentagon and the single edges are the geometries of triangles. The present section closely follows the construction in the original article.

Recall that the group $L_3(4)$ possesses three types of outer involutory automorphisms: *unitary* with centralizer $U_3(2) \cong 3^2 : Q_8$, *diagram* with centralizer $L_2(4) \cong A_5$ and *field* with centralizer $L_3(2)$.

We require the following properties of O'Nan's group $G = ON$ taken from the original discovery paper by M. O'Nan.

(ON1) There is a single class of involutions in G, and if σ is an involution in G then

$$C := C_G(\sigma) \cong 4 \cdot L_3(4) : 2,$$

where $C_0 := O^2(C)$ has index 2 in C. There is an element $\varepsilon \in C \setminus C_0$ which inverts $O_2(C)$ and performs a unitary automorphism of $C_0/O_2(C) \cong L_3(4)$.

(ON2) G has a single class of elements of order 3, and the centralizer of one of these elements ω is isomorphic to $3^2 \times A_6$ with $N_G(\langle\omega\rangle) \cong (3^2 : 2) \times A_6$.

(ON3) There is an involution $\tau \in \mathrm{Aut}(G) \setminus G$ (unique up to conjugation) such that $F := C_G(\tau)$ is Janko's first sporadic simple group, J_1, which is a maximal subgroup in G.

The required properties of $F = J_1$ are the following.

(J1) There is a single class of involutions in F, and, if σ is an involution in F, then

$$C_F(\sigma) \cong 2 \times A_5$$

is a maximal subgroup in F. There are no elements of order 4 in F.

(J2) F has a single class of elements of order 3, and the centralizer of one of these elements ω is isomorphic to $3 \times D_{10}$, with $N_G(\langle\omega\rangle) \cong D_6 \times D_{10}$ being a maximal subgroup in F.

(J3) F contains a maximal subgroup L isomorphic to $L_2(11)$, on the cosets of which it acts distance-transitively, with the corresponding graph $\Gamma(J_1)$ having the following intersection diagram:

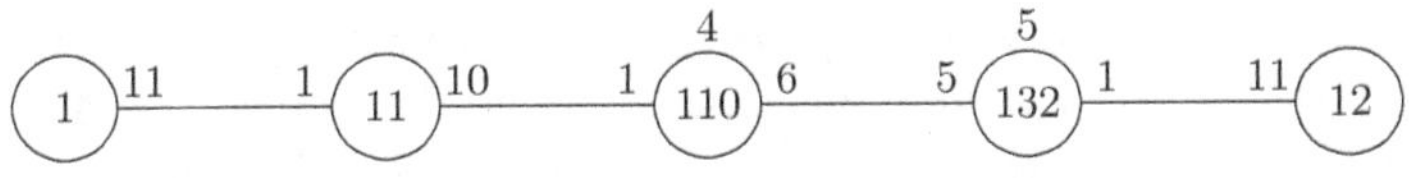

An involution $\sigma \in F$ fixes in $\Gamma(J_1)$ a Petersen subgraph on which $C_F(\sigma)$ induces the natural action of A_5. At the same time the set-wise stabilizer of an edge is also the centralizer of an involution. Thus the edges of $\Gamma(J_1)$ are canonically labelled by the involutions of F. Whenever two edges share a vertex, the corresponding involutions generate a dihedral group D_{10}. Thus every pair of intersecting edges is contained within a unique *special* pentagon, whose vertex-wise stabilizer is D_6 and whose set-wise stabilizer is $D_6 \times D_{10}$.

We proceed to the construction. Let τ be as in (ON3) and assume without loss of generality that τ centralizes σ as in (ON1). Then, by (J1),

$$H := C_G(\tau) \cap C_G(\sigma) \cong 2 \times A_5.$$

This shows that τ induces a diagram automorphism of $C_0/O_2(C) \cong L_3(4)$ and inverts $O_2(C)$ (note that, for a suitable choice of ε as in (ON1), the product $\tau\varepsilon$, being of order 8, induces a field automorphism of $C_0/O_2(C) \cong L_3(4)$ and centralizes $O_2(C)$. Since the centralizer of a diagram automorphism of $L_3(4)$ coincides with the stabilizer of an anti-flag in the associated projective plane of order 4 and this stabilizer is self-normalized in $L_3(4)$, we obtain the following:

$$N := N_G(H) \cong A_5 : D_8 \text{ with } C_N(O^2(H)) \cong 4 \text{ and } N_{\mathrm{Aut}\,(G)}(H) \cong A_5 : D_{16}.$$

Let Ω be the set of cosets of $F = C_G(\tau) \cong J_1$ in G and let α be the coset containing the identity. Then the structures of H and N presented above show that H stabilizes $4 = [N : H]$ cosets in Ω, say α, β_1, β_2 and β_3, so that, up to renumbering, the set-wise stabilizers of $\{\alpha, \beta_i\}$ in G are isomorphic to $2 \times S_5$ for $i = 1$ and 2 and to $4 \times A_5$ for $i = 3$. If the outer involution τ is chosen to normalize H, then it induces the permutation $(\alpha)(\beta_1\,\beta_2)(\beta_3)$.

Let D be a dihedral group of order 6 in F such that $N_F(D) \cong D_6 \times D_{10}$. By (J2) D is unique up to conjugation and by (ON2) $N_G(D) \cong D_6 \times A_6$. Therefore, if $\Omega(D)$ denotes the set of cosets in Ω stabilized by D, then

$$|\Omega(D)| = 36 = [A_6 \colon D_{10}],$$

and $N_G(D)$ acts on $\Omega(D)$ as A_6 acts on the cosets of D_{10}. Assuming without loss of generality that D commutes with σ, we have $D \leq O^2(H)$ and $\alpha, \beta_1, \beta_2, \beta_3 \in \Omega(D)$. The set $\Omega(D)$ viewed as an A_6-set can be described as follows. Let $\mathcal{A}_5^{(1)}$ and $\mathcal{A}_5^{(2)}$ be the two conjugacy classes of the A_5-subgroup in A_6. Then, for $A_5^{(1)} \in \mathcal{A}_5^{(1)}$ and $A_5^{(2)} \in \mathcal{A}_5^{(2)}$, we have $A_5^{(1)} \cap A_5^{(2)} \cong D_{10}$ and so

$\Omega(D)$ can be identified with the pairs of A_5-subgroups in A_6 taken from different conjugacy classes. From this description it follows that $A_5^{(j)} \in \mathcal{A}_5^{(j)}$ has an orbit $\Theta^{(j)}$ of length 6 in $\Omega(D)$ on which it acts doubly transitively, where $j = 1, 2$. Since the centralizer of an involution in A_6 is isomorphic to D_8, we observe that $N_F(D) \cong D_6 \times D_{10}$ has three orbits of length 5 on $\Omega(D)$ with set-wise stabilizers of pairs isomorphic to 2^2, 2^2 and 4. Thus, assuming that $A_5^{(1)}$ and $A_5^{(2)}$ intersect in the direct factor D_{10} of $N_F(D)$ and performing a renumbering, if necessary, we conclude that

$$\beta_1 \in \Theta^{(1)} \text{ and } \beta_2 \in \Theta^{(2)}.$$

Furthermore, if $X^{(1)} = D \times A_5^{(1)}$ and $X^{(2)} = D \times A_5^{(2)}$ are subgroups in $N_G(D) \cong D_6 \times A_6$ intersecting in $N_F(D) \cong D_6 \times D_{10}$ then the orbit of α under $X^{(i)}$ has length 6, containing β_i, and $X^{(i)}$ acts on this orbit doubly transitively, with D being the kernel.

Let Ξ be the smallest G-invariant graph on Ω in which α is adjacent to β_1. Then F acts on the set $\Xi(\alpha)$ of neighbours of α in Ξ as it acts on the set of its involutions via conjugation or, equivalently, as on the set of edges of the distance-transitive graph $\Gamma(J_1)$. Then, subject to the latter identification, $\Theta^{(1)}$ consists of α and the edges of a special pentagon in $\Gamma(J_1)$ and, by the double transitivity of the action of $X^{(1)}$, on $\Theta^{(1)}$, the subgraph of Ξ induced by $\Theta^{(1)}$ is complete. Let $v \in \Gamma(J_1)$ be a vertex in this special pentagon and $L \cong L_2(11)$ be the stabilizer of v on $F = J_1$. Then, by the above and the uniqueness feature of special pentagons, the set Ψ formed by α together with the 11 vertices in $\Xi(\alpha)$ corresponding to the edges of $\Gamma(J_1)$ containing v (in particular including β_1) is a 12-vertex complete subgraph in Ξ, intersecting $\Theta^{(1)}$ in three vertices: α, β_1 and some γ. Our next aim is to show that the set-wise stabilizer of Ψ in G acts (triply) transitively on Ψ.

Since $X^{(1)}$ acts on $\Theta^{(1)}$ as A_5 acts on the cosets of D_{10}, there is an element $g \in X^{(1)}$, which induces on $\Theta^{(1)}$ a permutation

$$g = (\alpha\, \beta_1)(\gamma) \ldots .$$

On the other hand, $A := G(\alpha) \cap G(\beta_1) \cap L$ acts transitively on the set $\Psi \setminus \{\alpha, \beta_1\}$ of size 10. Since A is characteristic in the stabilizer of α and β_1 in G (which is isomorphic to $A_5 \times 2$), the element g normalizes A. Therefore the image of $\Psi \setminus \{\alpha, \beta_1\}$ under g is also an orbit of A and, since γ is fixed by g, we may draw the crucial conclusion that $M := \langle L, g \rangle$ stabilizes Ψ. Since $L_2(11)$ acting on 11 points possesses a unique transitive extension, it is isomorphic to M_{11} and we have reached the ultimate conclusion of the section.

Proposition 9.1 *The Mathieu group M_{11} is contained in O'Nan's group.* □

If we define the graph Ψ using β_2 instead of β_1 we obtain a representative of another class of M_{11}-subgroups in ON. The two classes of M_{11}-subgroups in ON are fused in Aut (ON).

9.2 *K*3 Surfaces

A $K3$ surface S is defined as a complex or algebraic smooth minimal complete surface that is regular and has a trivial canonical bundle. The study of possible finite automorphism groups F of S was initiated by V. V. Nikulin,[5] who determined in particular the abelian automorphism groups of $K3$ surfaces. An automorphism group F of S is called *symplectic* if it fixes a nowhere-zero holomorphic 2-form ω on S. The classification of the finite groups which have symplectic actions on $K3$ surfaces[6] exhibited a remarkable connection with the Mathieu group M_{24}. Later[7] this connection received a conceptual explanation in terms of the Niemeier lattices (except for the Leech lattice). The lattice of type A_1^{24} is most closely related to the Mathieu group, since

$$A_1^{24} = \left\langle \frac{1}{2} \sum_{i=1}^{24} a_i e_i \mid a_i \in \mathbb{Z} \text{ and } (a_i) \bmod 2 \in \mathcal{C} \right\rangle,$$

where $\mathcal{C}$ is the Golay code and $\{e_i \mid 1 \le i \le 24\}$ is an orthogonal basis with $(e_i, e_i) = 2$. The automorphism group of this lattice is $2^{24} : M_{24}$.

Mukai's theorem claims that for a finite group F the following two conditions are equivalent:

(S) F possesses a faithful symplectic action on a $K3$ surface.

(M) F is isomorphic to a subgroup of M_{24} which acts on the 24-element point set P of the Steiner system with at least five orbits, and at least one of those orbits has length 1.

A crucial role in Mukai's proof was played by the following observation. If φ is a symplectic automorphism of order n of a $K3$ surface then, as was shown in Nikulin's paper, for $n \le 8$, the number f_n of fixed points of φ depends only on n and is known explicitly. On the other hand, for every $2 \le n \le 8$ the group M_{24} contains a unique class of elements of order n with no fixed-point free

[5] V. V. Nikulin, Konechnye gruppy avtomorfizmov kelerovykh poverkhostei tipa $K3$ [Finite groups of automorphisms of Kählerian surfaces of type $K3$], *Trudy Mosk. Mat. Ob.* [*Trans. Moscow Math. Soc.*] **38** (1980), 71–137.

[6] S. Mukai, Finite groups of automorphisms of $K3$ surfaces and the Mathieu group, *Invent. Math.* **94** (1988), 183–221.

[7] S. Kondo, Niemeier lattices, Mathieu groups, and finite groups of symplectic automorphisms of $K3$ surfaces, *Duke. Math. J.* **92** (1998), 593–603.

Table 9.1

n^0	F	$\|F\|$	Orbit lengths	Polarised $K3$-surfaces
1	$L_3(2)$	168	1,1,7,7,8 1,1,7,14	$X^3Y + Y^3Z + Z^3X + Y^4 = 0$ in $\mathbb{P}^3$
2	A_6	360	1,1,1,6,15 1,1,6,6,10	$\Sigma_1^6 X_i = \Sigma_1^6 X_i^2 = \Sigma_1^6 X_i^3 = 0$ in $\mathbb{P}^5$
3	S_5	120	1,1,2,5,15 1,2,5,6,10	$\Sigma_1^5 X_i = \Sigma_1^6 X_i^2 = \Sigma_1^5 X_i^3 = 0$ in $\mathbb{P}^5$
4	$2^4 : A_5$	960	1,1,1,5,16 1,1,1,1,20	$X^4 + Y^4 + Z^4 + T^4 + 12XYZT = 0$ in $\mathbb{P}^3$
5	$2^4 : S_4$	384	1,1,2,4,16	$X^4 + Y^4 + Z^4 + T^4 = 0$ in $\mathbb{P}^3$
6	$2^4 : (S_3 \times S_3)^+$	288	1,1,3,3,16	$X^2 + Y^2 + Z^2 = \sqrt{3}U^2$ $X^2 + \omega Y^2 + \omega^2 Z^2 = \sqrt{3}V^2$ in $\mathbb{P}^5$ $X^2 + \omega^2 Y^2 + \omega Z^2 = \sqrt{3}W^2$
7	$(Q_8 * Q_8) : S_3$	192	1,1,6,8,8	$X^4 + Y^4 + Z^4 + T^4 -$ $2\sqrt{-3}(X^2Y^2 + Z^2T^2) = 0$
8	$2^4 : D_{12}$	192	1,3,4,8,8	$X_1^2 + X_3^2 + X_5^2 = X_2^2 + X_4^2 + X_6^2$ $X_1^2 + X_4^2 = X_2^2 + X_5^2 = X_3^2 + X_6^2$ in $\mathbb{P}^5$
9	$3^2 : D_8$	72	1,2,6,6,9	$X_1^3 + X_2^3 + X_4^3 =$ $X_1X_2 + X_3X_4 + X_5^2 = 0$ in $\mathbb{P}^4$
10	$3^2 : Q_8$	72	1,1,1,9,12	Double cover of $\mathbb{P}^2$ with branch $X^6 + Y^6 + Z^6 -$ $10(X^3Y^3 + Y^3Z^3 + Z^3X^3) = 0$
11	$Q_8 : S_3$	48	1,3,4,8,8	Double cover of $\mathbb{P}^2$ with branch $XY(X^4 + Y^4) + Z^6 = 0$

action on P. If g_n is the permutation character of that element of order n on P, then f_n and g_n are the same and can be computed by using the following formula:

$$f_n = g_n = 24\left(n\prod_{p|n}\left(1+\frac{1}{p}\right)\right)^{-1}.$$

It is clear that the class of groups satisfying conditions (S) and (M) is closed under taking subgroups, thus it is essential to describe the maximal subgroups subject to these properties. There are 11 groups up to isomorphism, which are listed in Mukai's original paper and in his supplement to Kondo's paper. The information on these groups is given in Table 9.1. Notice that for the first group, $L_3(2)$, the subspace spanned by X, Y and Z supports the complex version of the three-dimensional representation as in Section 6.2.

It would be attractive to deduce directly from the properties of the Mathieu group M_{24} and its action on the Golay code that the 11 groups in Table 9.1 are exactly the maximal subgroups of M_{24} satisfying the condition (M), although it does not appear to be particularly easy to do so, thus inspiring in the reader great admiration for Mukai's classification of such groups. It is most remarkable that the proof was achieved before any conceptual explanation had been developed.

For the conceptual explanation, S. Kondo considered the sublattice L_F of $H^2(S, \mathbb{Z})$ for a finite group F acting by symplectic automorphisms on a $K3$ surface S, which is orthogonal to the F-invariant sublattice. It had already been shown by V. V. Nikulin that L_F is an even, negative definite lattice with rank less than or equal to 19, and that it contains no (-2)-elements. Kondo showed that L_F can be primitively embedded into the lattice A_1^{24}, which shows that F acts on the Dynkin diagram of A_1^{24} (which corresponds to the 24 root subspaces), stabilizing at least one vertex. This enabled him to conclude that F belongs to the class (M).

9.3 Mathieu Moonshine

A new epoch in the history of the Mathieu group M_{24} started when T. Eguchi, H. Ooguri and Y. Tachikawa[8] observed that the dimensions of some representations of M_{24} (the smallest being the 90-dimensional sum of two algebraically conjugate 45-representations constructed in Chapter 6) are multiplicities of superconformal algebra characters in the $K3$ elliptic genus. The new area which goes under the name of *Mathieu moonshine* is still pretty much in the making. The original Eguchi–Ooguri–Tachikawa conjecture has recently been proved by T. Gannon.[9] On the way to this proof many more observations for which there are as yet no conceptual explanations were made. The best we can do here is to encourage the reader to follow the ongoing research on this fascinating topic (as the author does), hoping to join the game when the time comes.

[8] T. Eguchi, H. Ooguri and Y. Tachikawa, Notes on the $K3$ surface and the Mathieu group M_{24}. *Exper. Math.* **20** (2011), 91–96.

[9] T. Gannon, Much ado about Mathieu, *Adv. Math.* **301** (2016), 322–358.

10

Locally Projective Graphs and Amalgams

Many graphs constructed and discussed in this book fall within the class of *locally projective graphs* (with respect to a relevant group) according to the definition given in the first section of this chapter. Since some of these graphs were discussed in detail, it appears reasonable to present here a complete list of known examples and outline a project for the complete classification of the corresponding vertex-line amalgams.

10.1 Definitions and Preliminaries

We start with our principal definitions.

Definition 10.1 *Let Γ be a connected graph and let G be a group of automorphisms of Γ. Then Γ is said to be* locally projective *of type (n, α), where $n, \alpha \in \mathbb{N}$ with respect to the action of G, whenever the following conditions hold:*

(i) *G acts vertex- and edge-transitively on Γ;*
(ii) *there is a family $\mathcal{L}$ of complete subgraphs in Γ (called* lines*) having α vertices each, such that (a) $\mathcal{L}$ is preserved by G and (b) every edge of Γ is contained in a unique line from $\mathcal{L}$;*
(iv) *every vertex x of Γ is contained in exactly $(2^n - 1)$ lines, and the stabilizer $G(x)$ of x in G induces on this $(2^n - 1)$-set of lines the natural doubly transitive action of the group $L_n(2)$ as on the set of points of the corresponding projective $GF(2)$-geometry π_x;*
(v) *the stabilizer in G of a line acts doubly transitively on the vertex-set of the line;*
(vi) *if $\alpha = 2$ then G is not transitive on 3-paths in Γ and, whenever $\{x, y\}$ is an edge, an element swapping x and y induces a collineation (rather than correlation) between the residue of y in π_x and the residue of x in π_y.*

We assume that α is either 2 or 3. If $\alpha = 2$ then $\mathcal{L}$ is the edge-set of Γ. Suppose that $\alpha = 3$. Then $\mathcal{L}$ is a family of triangles in Γ; the stabilizer $G(l)$ of a line-triangle l induces on its vertices the symmetric group $S_3 \cong L_2(2)$. Since by (ii) (b) any two lines intersect in at most one vertex, the valency of Γ is $2 \cdot (2^n - 1)$. The $GF(2)$-vector spaces whose non-zero vectors are indexed by the lines passing through x will be called the *natural module* of the group $L_n(2)$ induced by $G(x)$ on the set of these lines.

Definition 10.2 *Let Γ be a graph which is locally projective of type (n, α) with respect to a group G. Let x be a vertex of Γ and let l be a line containing x. Then the amalgam*

$$\mathcal{A} = \{G(x), G(l)\}$$

is said to be a locally projective amalgam *of type (n, α).*

For some small pairs (n, α) the locally projective amalgams have already been classified. D. Z. Djoković and G. L. Miller[1] used Tutte's classical theorem[2] to prove the following theorem (the seventh amalgam (Z_3, Z_2) from their theorem does not fall within the locally projective class).

Theorem 10.3 *There are exactly six locally projective amalgams of type* (2, 2) *as follows:*

$$\{S_3, 2^2\},\ \{S_3, Z_4\},\ \{D_{12}, D_8\},\ \{S_4, D_{16}\},\ \{S_4, Q_8 : 2\},\ \{S_4 \times 2, SD_{32}\}.$$

□

The locally projective amalgams of type (2, 3) are Goldschmidt's amalgams in Table 2.1 except for a few of the small ones (namely G_1, G_1^2, G_2 and G_2^3). There is a relationship between Djoković–Miller amalgams and 'symmetric' Goldschmidt amalgams. In fact, if G acts vertex- and edge-transitively on a cubic graph, then the stabilizers of two vertices on an edge form a Goldschmidt amalgam (P_1, P_2) which possesses an outer automorphism ρ which permutes P_1 and P_2 and normalizes $P_1 \cap P_2$ (ρ is induced by an element of G which flips the vertices on the edge). On the other hand, if (P_1, P_2) is a Goldschmidt amalgam which possesses an automorphism ρ as above, then $(P_1, \langle P_1 \cap P_2, \rho\rangle)$ is a Djoković–Miller amalgam. Different choices of ρ might lead to different amalgams. For example, the Goldschmidt amalgam $G_3 = \{S_4, S_4\}$ is realized in A_6,

[1] D. Z. Djoković and G. L. Miller, Regular groups of automorphisms of cubic graphs, *J. Combin. Theory* (B) **29** (1980), 195–230.

[2] W. T. Tutte, A family of cubical graphs, *Proc. Camb. Phil. Soc.* **43** (1947), 459–474.

Table 10.1

n	$\mathcal{A}$	$G(x)/G_1(x)$	V_2	V_3	V_4	V_5	Some completions
3	$\mathcal{A}_3^{(1)}$	$L_3(2)$	2^3				–
	$\mathcal{A}_3^{(2)}$	$L_3(2)$	2^3				M_{22}
	$\mathcal{A}_3^{(3)}$	$L_3(2)$	2^3				–
	$\mathcal{A}_3^{(4)}$	$L_3(2)$	2^3	2			$(S_8 \wr 2)^+$
	$\mathcal{A}_3^{(5)}$	$L_3(2)$	2^3	2			$\mathrm{Aut}(M_{22})$
4	$\mathcal{A}_4^{(1)}$	$L_4(2)$					M_{23}
	$\mathcal{A}_4^{(2)}$	$L_4(2)$	2^6				A_{64}
	$\mathcal{A}_4^{(3)}$	$L_4(2)$	2^6	2^4	2		Co_2
	$\mathcal{A}_4^{(4)}$	$L_4(2)$	2^6	2^4	2^4		J_4
	$\mathcal{A}_4^{(5)}$	$L_4(2)$	2^6	2^4	2^4		A_{256}
5	$\mathcal{A}_5^{(1)}$	$L_5(2)$	2^{10}				J_4
	$\mathcal{A}_5^{(2)}$	$L_5(2)$	2^{10}	2^{10}	2^5	2^5	BM

whereas the amalgams $\{S_4, D_{16}\}$ and $\{S_4, Q_8 : 2\}$ are realized in $PGL_2(9)$ and M_{10}, both being extensions of A_6 by outer automorphisms.

The locally projective amalgams of type $(n, 2)$ for all $n \geq 3$ were classified by S. V. Shpectorov and the present author,[3] making use of a fundamental result obtained by V. I. Trofimov,[4] The classification is given in the following theorem, where $G_i(x)$ denotes the vertex-wise stabilizer in G of the ball of radius i in Γ centred at x and $V_i = G_i(x)/G_{i-1}(x)$.

Theorem 10.4 *Let G be a group acting locally projectively on a graph Γ of type $(n, 2)$ for some $n \geq 3$, and let $\mathcal{A} = \{G(x), G(l)\}$ be the corresponding locally projective amalgam. Then one of the following three possibilities holds:*

(i) *$\mathcal{A}$ is isomorphic to the locally projective amalgam associated with the natural action of the affine group $AGL_n(2)$ on the vector-set of the corresponding n-dimensional $GF(2)$-space;*
(ii) *$\mathcal{A}$ is isomorphic to the locally projective amalgam associated with the natural action of the orthogonal group $O_{2n}^+(2)$ on the corresponding dual polar space graph;*

[3] A. A. Ivanov and S. V. Shpectorov, Amalgams determined by locally projective actions, *Nagoya Math. J.* **176** (2004), 19–98.
[4] V. I. Trofimov, Vertex stabilizers of locally projective groups of automorphisms of graphs: A summary, in *Groups, Combinatorics and Geometry*, ed. A. A. Ivanov, M. W. Liebeck and J. Saxl, World Scientific, River Edge, NJ, 2003, pp. 313–326

(iii) *$\mathcal{A}$ is one of the 12 exceptional amalgams in Table 10.1 represented by their members $G(x)$, where all the quotients V_i are elementary abelian 2-groups.* □

10.2 Some $(n, 3)$-Examples

We are exploring the possibility of extending this classification to the projective amalgams of type $(n, 3)$. We expect that the classification will be achieved through reduction to the $(n, 2)$-case through so-called densely embedded subgraphs, and we start by listing the known examples (including numerous ones that have already been seen in this book).

We require a further piece of notation. For a group G acting locally projectively on a graph Γ of type $(n, 3)$, we denote by $G_{i\frac{1}{2}}(x)$ the largest subgroup in $G_i(x)$ which stabilizes every line at distance at most i from x in Γ and let $V_0 = G_{\frac{1}{2}}(x)/G_1(x)$.

The entry 11a in Table 10.2 contains the largest locally projective automorphism group of the Hamming graph $H(n, 3)$. A variety of smaller subgroups can be obtained along the following lines: (a) construct $G(x)$ as an extension by $L_n(2)$ of a submodule in the $GF(2)$-permutation module of $L_n(2)$ on the point set; (b) consider the stabilizer in $G(x)$ of a line and denoted it by $G(x) \cap G(l)$; (c) if possible choose a subgroup X of index 2 in $G(x) \cap G(l)$; and (d) induce the one-dimensional $GF(3)$-character χ of $G(x) \cap G(l)$ with kernel X to the whole of $G(x)$; (e) then $G \cong 3^n.(G(x) \cap G(l)$ is the group associated with the induced action.

10.3 Densely Embedded and Geometric Subgraphs

When dealing with locally projective amalgams we usually assume that the graph Γ is a bipartite half of the coset graph of the universal completion of the corresponding amalgam. By the universality property, the coset graph is a tree with the vertices in one half having valency $(2^n - 1)$ and the vertices in the other half (the lines) having valency 3. The problem we consider is a pushing-up type problem for $L_n(2)$ and it is widely open, with only a few special cases having been settled thus far.[5]

We start by defining a *densely embedded subgraph* in the locally projective graph of type $(n, 3)$.

[5] C. Parker and P. Rowley, On the non-commuting case for $(S_3, L_3(2))$-amalgams, *J. Algebra* **181** (1996), 267–285.

Table 10.2

#	$G(x)/G_{\frac{1}{2}}(x)$	α	V_0	V_2	V_3	V_4	V_5	G	Reference
1	$L_3(2)$	3						A_7	Section 2.5
2a	$L_3(2)$	3	2^3	2^3	2			M_{24} or He	Section 7.7
2b	$L_3(2)$	2		2^3	2			$\mathrm{Aut}(M_{22})$	Section 5.2
3a	$L_3(2)$	3	2^3	2^3	2			A_{16}	Giudici *et al.*[a]
3b	$L_3(2)$	2		2^3	2			$(S_8 \wr 2)^+$	Table 10.1
4	$L_3(2)$	3		2^3				$G_2(3)$	Cooperstein[b]
5	$L_3(2)$	3	2	2^3				$G_2(3).2$	Cooperstein[b]
6a	$L_4(2)$	3	2^4					M_{24}	Section 1.2
6b	$L_4(2)$	2						M_{23}	Section 5.1
7a	$L_4(2)$	3	2^4	2^6	2^4	2		Co_1	Ronan and Stroth[c]
7b	$L_4(2)$	2		2^6	2^4	2		Co_2	Ronan and Stroth[c]
8a	$L_5(2)$	3	2^5	2^{10}	2^{10}	2^5	2^6	M	Ronan and Stroth[c]
8b	$L_5(2)$	2		2^{10}	2^{10}	2^5	2^5	BM	Ronan and Stroth[c]
9a	$L_n(2)$	3	2^n					$L_{n+1}(2)$	Section 2.6 for $n = 3$
9b	$L_n(2)$	2						$2^n : L_n(2)$	Theorem 10.4 (i)
10a	$L_n(2)$	3	2^n	$2^{n(n-1)/2}$				$Sp_{2n}(2)$	Section 4.6 for $n = 4$
10b	$L_n(2)$	2		$2^{n(n-1)/2}$				$O^+_{2n}(2)$	Theorem 10.4 (ii)
11a	$L_n(2)$	3	2^n					$S_3 \wr L_n(2)$	Hamming graph
11b	$L_n(2)$	2						$2^n : L_n(2)$	Binary cube

[a] M. Giudici, A. A. Ivanov, L. Morgan and C. E. Praeger, A characterisation of weakly locally projective amalgams related to A_{16} and the sporadic simple groups M_{24} and He, *J. Algebra* **460** (2016), 340–365.
[b] B. N. Cooperstain, A finite flag-transitive geometry of extended G_2-type, *Europ. J. Combin.* **10** (1989), 313–318.
[c] M. A. Ronan and G. Stroth, Minimal parabolic geometries for the sporadic groups, *Europ. J. Combin.* **5** (1984), 59–91.

Definition 10.5 *Suppose that G acts locally projectively on Γ of type $(n,3)$ for $n \geq 3$, and let Δ be a connected subgraph in Γ. Then Δ is said to be* densely embedded *in Γ if the following conditions hold:*

(i) *the subgroup H of G which stabilizes Δ as a whole induces on it a locally projective action of type $(n,2)$, possibly with a non-trivial kernel;*
(ii) *if $x \in \Delta$ then $H(x)$ contains $G_1(x)$ and $H(x)/H_1(x)$ is an $L_n(2)$-complement to $G_{\frac{1}{2}}(x)/G_1(x)$ in $G(x)/G_1(x)$.*

It is implicit in Definition 10.5 (ii) that a densely embedded subgraph exists only if $G(x)/G_1(x)$ splits over $G_{\frac{1}{2}}(x)/G_1(x)$. In fact, densely embedded subgraphs exist quite often: in Table 10.2 the subgraph number Nb is densely embedded into the subgraph Na for

$$N = 2, 3, 6, 7, 8, 9, 10, 11.$$

This observation served as a starting point for our classification project, which is still in progress. More specifically, we observed that, since both M_{24}- and He-examples correspond to the same locally projective amalgam 2a, and M_{24} contains a densely embedded subgraph (stabilized by $\mathrm{Aut}(M_{22})$), the universal cover of the Held graph must contain a densely embedded subgraph. In fact, when the universal cover is folded onto the Held graph, the densely embedded subgraph folds onto the whole of the Held graph, so it is hard to see its traces in the He-graph.

We are aiming to turn the existence of densely embedded subgraphs into a theorem (subject to certain constraints on the vertex stabilizer $G(x)$) and to deal with the exceptional cases by *ad hoc* methods. Notice that in Table 10.2 there is only one instance in which the action on Δ of its stabilizer in G is unfaithful (with kernel of order 2): G is the Monster, 8a, and H is the double cover of the Baby Monster, 8b.

Our procedure for constructing densely embedded subgraphs is simply minded and consists of the following steps:

(DE1) assuming that $G(x)$ splits over $G_{\frac{1}{2}}(x)$ we take the full pre-image in $G(x)$ of an $L_n(2)$-complement, and denote it by $H(x)$ (although H is not yet known);

(DE2) intersect $H(x)$ with the vertex-wise stabilizer of $l = \{x, y, z\}$ to obtain the subgroup $H(x, y, z)$;

(DE3) search for elements $\sigma \in G(l)$ which normalize $H(x, y, z)$ and swap x either with y or with z;

(DE4) if and when the required σ has been found, we put $H = \langle H(x), \sigma \rangle$ and define Δ to be the subgraph on the set of images of x under H.

Next we introduce *geometric subgraphs*. Recall that every vertex x is equipped with a $GF(2)$-space π_x, whose points are the lines containing x.

Definition 10.6 *A connected subgraph Ξ in Γ is said to be* geometric *at level k, where $1 \le k \le n-2$, whenever together with an edge it always contains the line on this edge, and the following conditions hold:*

(i) *if $x \in \Xi$, then the set of neighbours $\Xi(x)$ of x in Ξ is a k-dimensional subspace in π_x and the stabilizer of $\Xi(x)$ in $G(x)$ stabilizes Ξ;*

(ii) *the stabilizer F of Ξ in G acts on Ξ locally projectively with type $(k, 3)$ (probably with a non-trivial kernel).*

It is clear that the geometric subgraphs at level 1 are just the lines, whereas those at level 2 (called *planes*) are of valency 6 and realize a locally projective action of a faithful completion of a Goldschmidt amalgam. In general, geometric subgraphs might not exist. For instance, in Sections 4.6 and 4.7 we saw that the octad graph does not contain geometric subgraphs at level 3. The geometric subgraphs at level 4 do not exist in the locally projective graph of type $(5, 2)$ associated with the amalgam $\mathcal{A}_5^{(2)}$ in Table 10.1.

In Section 10.5 we will show that under some assumptions geometric subgraphs exist at all levels whenever Γ is coming from the coset graph of the universal completion of the corresponding locally projective amalgam.

10.4 The Thompson–Wielandt–Weiss Theorem

In this section we apply the following theorem[6] to demonstrate that in a locally projective action $G_1(x)$ is always a 2-group.

Theorem 10.7 *Let Λ be a connected, finite, undirected graph, $\{x, l\}$ be an edge of Λ and L be a subgroup of* $\mathrm{Aut}(\Lambda)$ *such that $L(v)^{\Lambda(v)}$ is primitive for $v = x$ and l. Let μ be the set of primes dividing the order of $L_1(x, l) := L_1(x) \cap L_1(l)$. Then either $|\mu| = 1$ or there exists a $p \in \mu$ such that for either $u = x$ and $v = l$ or $u = l$ and $v = x$, $O^p(L_1(x, l))) \leq L_2(u)$, $L_2(v)$ is a p-group and $L_2(v) = L_3(v)$.* □

The finiteness condition on Λ in the above theorem would not cause a problem, since any rank-2 amalgam is known to have a finite faithful completion. Apparently the proof of the theorem can be slightly adjusted to change 'finite' to 'locally finite'.

Theorem 10.8 *If we let G act on a locally projective graph of type $(n, 3)$ for $n \geq 3$, then $G_1(x)$ is a 2-group.*

Proof We apply Theorem 10.7 to the bipartite coset graph Λ of the completion $G = L$ of the corresponding locally projective amalgam. Since $G(l)^{\Lambda(l)} \cong S_3$ and $G(x)^{\Lambda(x)} \cong L_n(2)$ (both doubly transitive), the hypothesis in the theorem holds. Suppose that $G_1(x)$ is not a 2-group and put $l = \{x, y, z\}$. Then $G_1(x)^{\Gamma(y)}$ is divisible by more than one prime. Notice that in this case

$$G(y)/\langle O_2(G(y), G_1(y)\rangle \cong L_{n-1}(2)'.$$

[6] R. Weiss, Elations of graphs, *Acta Math. Acad. Scient. Hungary* **34** (1979), 101–103.

Consider the exceptional configurations in Theorem 10.7. Notice that

$$L_1(x,l) = G(x,y,z) \cap G_{\frac{1}{2}}(x),\ L_2(x) = G_1(x),\ L_2(l) = \bigcap_{a \in l} G_{\frac{1}{2}}(a),\ L_3(l) = \bigcap_{a \in l} G_1(a).$$

First suppose that $u = x$ and $v = l$. Then, to justify the inclusion $O^p(L_1(x,l)) \leq G_1(x)$, we must have $p = 2$, but, since $G_1(x)^{\Gamma(y)}$ is divisible by more than one prime, the inclusion $O^2(L_1(x,l)) \leq G_1(x)$ fails. The second exceptional configuration can be dealt with in a similar manner. □

10.5 Large Goldschmidt Amalgams

We require some detailed information on Goldschmidt amalgams whose Borel subgroup B has order greater than 2^4. It follows immediately from Table 2.1 that there are five such amalgams:

$$G_3^1,\ G_4,\ G_4^1,\ G_5,\ G_5^1,$$

where G_4 and G_5 are 2-perfect, while the others are not. We start by summarizing the content of pp. 389–391 of Goldschmidt's paper,[7] omitting some proofs but adding some details needed for the future exposition.These details are summarized in Table 10.3. In this section B will denote the *Borel subgroup* of the amalgam $\{P_1, P_2\}$ under consideration, that is, $B = P_1 \cap P_2$.

The definition of the amalgams G_4, G_4^1, G_5 and G_5^1 is based on subgroups of the automorphism group of a direct product J of two cyclic groups of order 4 with generators a and b. The automorphisms which will play important roles are the following ones:

$$\begin{array}{lll} s: & a \mapsto b, & b \mapsto a, \\ t: & a \mapsto a^{-1}, & b \mapsto b^{-1}, \\ x: & a \mapsto b, & b \mapsto a^{-1}b^{-1}, \\ u: & a \mapsto ab^2, & b \mapsto ba^2, \\ v: & a \mapsto ab^2, & b \mapsto b^{-1}a^2. \end{array}$$

Then $A = \langle u, x, s, v\rangle$ is $\mathrm{Aut}(J)$, which is isomorphic to the semidirect product of $A_0 = \langle u, u^x, v, y\rangle \cong 2^4$ and $\langle x, s\rangle \cong S_3$.

10.5.1 G_4 and G_4^1

Let $P_1 = \langle a, b, x, s\rangle$ be the semidirect product of $\langle a, b\rangle \cong Z_4 \times Z_4$ and S_3. Put $B = \langle a, b, x\rangle$ and define $Q_1 = \langle abs, sa^2\rangle$ and $R = \langle Q_1, ab\rangle$. Then $Q_1 \cong Q_8$, $R \cong Q_8 \star Z_4$, $B = \langle R, sa\rangle$. Define

[7] D. M. Goldschmidt, Automorphisms of trivalent graphs, *Ann. Math.* **111** (1980), 377–406.

$$P_2 = \langle B, y \mid y^3 = 1, (abs)^y = a^2s, (sa^2)^y = ab^{-1}, y^{sa} = y^{-1}\rangle.$$

Then $G_4 = \{P_1, P_2\}$, and $G_4^{\natural}$ is obtained by adjoining to G_4 the automorphism t which inverts all the elements of J.

10.5.2 G_5 and $G_5^{\natural}$

Let $P_1 = \langle a, b, x, s, t\rangle$ be a semidirect product of $Z_4 \times Z_4$ by D_{12}, and let $B = \langle a, b, s, t\rangle$. As in the previous subsection, set $Q_1 = \langle abs, sa^2\rangle$. Also set $Q_2 = \langle ab, tsa^2\rangle$. We have already verified that $[Q_1, Q_2] = 1$ and that $(sa)^{tsa^2} = (sa)^3$. Since $ab = sa^2$, we have $(ab)^{tsa^2} = (ab)^{-1}$, so $Q_2 \cong Q_8$ and $Q_1Q_2 \cong Q_8 \star Q_8$. We have $B = \langle Q_1Q_2, bt\rangle$ where $(bt)^2 = 1$ and

$$(abs)^{bt} = a^2s, \ (tysa^2)^{bt} = (tb^2b^{-1}asa^2)^t = tab^{-1}a^{-1}a^2 = (tsa^2)(ab)^{-1}.$$

Define

$$P_2 = \langle B, y \mid y^3 = 1, (abs)^y = ab^{-1}, (sa^2)^y = abs, \\ (tsa^2)^y = ab, (ab)^y = tsa^{-1}b, y^{bt} = y^{-1}\rangle.$$

Then $Q_1Q_2 \cong Q_8 * Q_8$ is normal in P_2, $P_2/Q_1Q_2 \cong S_3$, and $G_5 = \{P_1, P_2\}$. A non-2-perfect extension $G_5^{\natural}$ is obtained by adjoining to B an involution which induces on J the automorphism υ.

10.5.3 Table of Chief Factors

Let $\Xi^{(0)}$ be the coset graph of a (possibly universal) faithful completion X of one of the five amalgams:

$$\mathcal{A} \in \{G_3^{\natural}, G_4, G_4^{\natural}, G_5, G_5^{\natural}\}.$$

Consider a connected component Ξ of the distance-2 graph of $\Xi^{(0)}$. Then X acts on Ξ locally projectively of type (2, 3), where the vertex stabilizer $X(u)$ is the relevant parabolic, that is $P_i(\mathcal{A})$, where $i = 1$ or 2 and $\mathcal{A}$ is from the above list of five amalgams. In Table 10.3 we summarize the information on the quotients

$$M_\alpha^\beta := X_\alpha(u)/X_\beta(u).$$

Table 10.3 can be justified along the following lines. One can label the vertices in one part of the coset graph by the central involutions of their stabilizers and those in the other part by the normal 2^2-subgroups in their stabilizers. As long as we are not too far away from the standard edge

$$\{a^2b^2, \langle a^2, b^2, a^2b^2\rangle\},$$

Table 10.3

$F(x)$	Structure	$\|B\|$	$M_0^{\frac{1}{2}}$	$M_{\frac{1}{2}}^1$	M_0^1	M_1^2	M_2^3
$P_1(G_3^1)$	$(S_4 \times 2)$	2^4	2^2	S_3	S_4	2	
$P_1(G_4)$	$(Q_8 * Z_4).S_3$	2^5	2^2	S_3	S_4	Z_4	
$P_2(G_4)$	$(Z_4 \times Z_4).S_3$	2^5	2^2	S_3	S_4	2^2	
$P_1(G_4^1)$	$(Q_8 * Q_8)^1.S_3$	2^6	2^3	S_3	$S_4 \times 2$	Z_4	
$P_2(G_4^1)$	$(Z_4 \times Z_4).D_{12}$	2^6	2^2	S_3	S_4	2^3	
$P_1(G_5)$	$(Q_8 * Q_3)^2.S_3$	2^6	2^2	S_3	S_4	2^2	2
$P_2(G_5)$	$(Z_4 \times Z_4).D_{12}$	2^6	2^2	S_3	S_4	2^3	
$P_1(G_5^1)$	$(Q_8 * Q_8)^2.D_{12}$	2^7	2^2	S_3	S_4	2^3	2
$P_2(G_5^1)$	$(Z_4 \times Z_4).(S_3 : D_8)$	2^7	2^3	S_3	$S_4 \times 2$	2^3	

the labelling is not ambiguous. By making use of Goldschmidt's presentations for the amalgam, it is relatively easy to label all the vertices at distance at most 2 from the standard edge. Then the action of the parabolics P_1 and P_2 becomes explicit and all the required information can be checked.

Another approach is to perform the calculations inside the completions $G_2(2)'$ and M_{12} of the amalgams G_4 and G_5. In the case of $G_2(2)'$ this is the corresponding generalized hexagon. In the case of M_{12} it is the graph on 4-subsets and 'special' $(4+4+4)$-partitions of the 12-set on which M_{12} acts. The following information, Lemmas 10.9 and 10.10, on involutions in $G_2(2)'$ and M_{12} proved to be useful and almost sufficient to recover the entries of Table 10.3. Notice that $P_1(G_3^1) \cong P_2(G_3^1)$.

Lemma 10.9 *The group $G_2(2)'$ contains a single class of involutions and a given involution commutes with seven involutions (including itself). An involution stabilizes just three lines of the $G_2(2)'$-hexagon.* □

By the above lemma, when $G_2(2)'$ acts on a graph Γ locally projectively with $G(x) = P_1(G_4) \cong (Q_8 * Z_4).S_3$, the central involution τ of $G(x)$ stabilizes vertex-wise $\{x\}$ and $\Gamma(x)$, stabilizing the three lines containing x and acting fixed-point freely on the lines at distance 1 from x.

Lemma 10.10 *The group M_{12} contains two classes of involutions, $2A$ and $2B$, with the central involution of $P_1(G_5)$ being a $2B$-involution. The centralizer of a $2B$-involution contains 24 involutions of type $2A$ and 24 of type $2B$. An involution of type $2A$ stabilizes 15 special $(4+4+4)$-partitions of the 12-set on which M_{12} acts naturally.* □

10.6 Geometric Subgraphs Exist in the Universal Cover

As above, let G act locally projectively on a graph Γ of type $(n, 3)$, where $n \geq 3$. Let $x \in \Gamma$ be a vertex, and let Δ_x be a subspace of dimension k, $2 \leq k \leq n - 1$, in the space π_x associated with x containing the line $l = \{x, y, z\}$. Then $G(x) \cap G(l)$ induces the full linear group $L_{n-1}(2)$ on the set of subspaces of π_x containing l. Under this action different subspaces have different stabilizers. In particular, Δ_x is the *unique* k-subspace in π_x that is stabilized by $G(x) \cap G(\Delta_x)$. In this section we prove the following theorem.

Theorem 10.11 *Let G act locally projectively on a graph Γ of type $(n, 3)$, where Γ is a bipartite part of the coset graph of the universal completion of the amalgam $\{G(x), G(l)\}$. Then Γ contains at least one G-orbit of geometric subgraphs at level k for every $2 \leq k \leq n - 1$.*

The proof will be divided into considerations of two cases specified by the validity of the following two conditions.

(I) $L_{n-1}(2)$ does not contain subgroups of index 2.
(II) $G_{\frac{1}{2}}(x)$ is not equal to $G_1(x)$ and it contains an element which swaps y with z, while acting trivially on the set of π_x-subspaces passing through l.

The following lemma is well known and easy to prove.

Lemma 10.12 *If neither condition* (I) *nor condition* (II) *holds then the action of $G(x, y, z)$ on the set of π_x-subspaces containing l is isomorphic to $L_{n-1}(2)$ and the uniqueness assertion from the paragraph before Theorem* 10.11 *holds. Furthermore, in this case $G(x)/G_1(x) \cong L_3(2)$ acts on $\Gamma(x)$ as on the cosets of A_4.*

Proof of Theorem 10.11 Suppose that the conditions of Lemma 10.12 hold. Then

$$X(x) := G(x, y, z) \cap G(\Delta_x)$$

stabilizes a unique k-subspace Δ_y in π_y and a unique such subspace Δ_z in π_z. By the obvious symmetry $N_{G(l)}(X(x))$ acts transitively on the points of l, permuting around the three subspaces. Then put

$$X = \langle N_{G(l)}(G(x, y, z)) \cap X(x), X(x) \rangle,$$

and define Ξ to be the subgraph induced by the images of x under X. □

The situation described in Lemma 10.12 (ii) is realized in the A_7-example, number 1 in Table 10.2, and in Cooperstein's graph, number 4. In fact, geometric subgraphs exist in both cases. For the A_7-example they are planes of the underlying projective geometry, whereas in the other example there are in fact three G-orbits of planes, such that representatives of two of the orbits are isomorphic to point graphs of the $G_2(2)$-generalized hexagons (realizing the amalgam G_4^1), while the representatives of the third orbit realize the amalgam G_5. The outer automorphism of the Cooperstein geometry permutes the former two orbits of planes and (when lifted to an automorphism of the universal completion) stabilizes the third orbit of planes, causing them to realize the G_5^1-amalgam. It was Chris Parker[8] who explained this remarkable phenomenon to me.

Thus the universal cover graph of the locally projective amalgam of Cooperstein's geometry of $G_2(3)$ contains *three* classes of geometric subgraphs: two of them can be folded onto the $G_2(2)'$-geometry, whereas the third one can be folded onto the geometry of M_{12} (which is the minimal completion of Goldschmidt's G_5-amalgam).

10.6.1 What If $G_{\frac{1}{2}}(x) = 1$?

Because of the vertex- and edge-transitivity, $L_n(2)$ acts transitively on the $2(2^n - 1)$ vertices in $\Gamma(x)$ so that the induced action on the lines containing x is the natural doubly transitive action of degree $(2^n - 1)$ on the cosets of the maximal parabolic subgroup $2^{(n-1)} : L_{n-1}(2)$. The latter group contains an index-2 subgroup only if $n = 3$ and $2^2 : L_2(2) \cong S_4$. Thus we have the following lemmas.

Lemma 10.13 *If* $G_{\frac{1}{2}}(x) = 1$ *we have* $n = 3$ *and* $G(x)/G_1(x) \cong L_3(2)$ *acts on* $\Gamma(x)$ *as on the cosets of an* A_4*-subgroup.* □

Lemma 10.14 *As in the hypothesis of Lemma* 10.13 *suppose that* $G_1(x) = 1$. *Then* $\{G(x), G(l)\}$ *is isomorphic to amalgam number* 1 *in Table 10.2.*

Proof Since $G(x) \cong L_3(2)$, it is easy to deduce that $G(l) \cong (S_3 \times S_4)^+$ and to reconstruct the whole amalgam uniquely up to isomorphism. □

Notice that besides A_7 the amalgam $\{L_3(2), (S_3 \times S_4)^+\}$ from A_7 is in the group $3^7 : L_3(2)$ acting locally projectively on the Hamming graph $\mathrm{Ham}(7, 3)$.

[8] C. W. Parker, private communication, 2017.

10.6.2 Trinity

In this subsection we handle the configurations (G, Γ) falling under the following hypothesis.

Hypothesis 10.15 *The following conditions hold:*

(i) *G acts locally projectively on Γ of type $(3, 3)$, and, unless stated otherwise, G is the universal completion of the corresponding locally projective amalgam, so that Γ has a tree-like structure with seven disjoint triangles passing through a vertex and there are no other proper cycles;*
(ii) *for a vertex x we have $G_{\frac{1}{2}}(x) = G_1(x)$, so that $G(x)$ acts on $\Gamma(x)$ as $L_3(2)$ acts on the 14 cosets of its A_4-subgroup, and a block of imprimitivity, say $\{y, z\}$, of this action together with x is a line;*
(iii) $G_1(x) \neq 1$.

We proceed by stating implicitly some direct consequences of Hypothesis 10.15.

Lemma 10.16 *If Hypothesis* 10.15 *holds then*

(i) *$G(x, y, z)$ induces on $\Gamma(x)$ an action of A_4, while $G(x, l)$ acts as S_4;*
(ii) *$G_1(x)$ induces on $\Gamma(y)$ an elementary abelian group of order 2^2 on which $G(x, y, z)$ induces an action of order 3;*
(iii) *the order of $G(x)$ is $2^m \cdot 3 \cdot 7$ for some $m \geq 6$ and $G_1(x)$ involves a non-trivial $G(x)/G_1(x) \cong L_3(2)$-module over $GF(2)$.*

Proof The assertion (i) is immediately evident. Since $G_1(x)$ is a non-trivial 2-group by Hypothesis 10.15 (iii) and Theorem 10.8, (ii) follows. The group $L_3(2)$ is known to have just four irreducible $GF(2)$-modules, namely the trivial module, the natural and dual natural modules of dimension 3, and the Steinberg module of dimension 8, therefore leading to the bound as in (iii). □

Lemma 10.17 *Let T be a Sylow 3-subgroup of $G(l)$. Then T is elementary abelian of order 9. Furthermore, if $\tau \in G(z)$ is a 2-element which normalizes T and acts on l as the transposition (x, y), then τ inverts every element of T.*

Proof Let $S^{(0)}$ be a Sylow 2-subgroup of $G(x, y, z)$. By Lemma 10.16, $Q := O_2(G(x, y, z))$ induces on $\Gamma(l) := \bigcup_{u \in l} \Gamma(u) \setminus l$ an elementary abelian 2-group of order at most 2^6 on which $S^{(0)}$ acts fixed-point freely. Therefore, the action of T on $Q/G_1(l)$ is contained in $SL_3(4)$, which does not have elements of order 9. This implies the structure of T. Since $G(l)$ acts as S_3 on the points of l, the result follows, since τ inverts $S^{(0)}$ by Lemma 10.16. □

With T as in Lemma 10.17, let $S^{(0)}$, $S^{(1)}$, $S^{(2)}$ and $S^{(3)}$ constitute the complete set of order-3 subgroups in T, so that $S^{(0)}$ fixes l vertex-wise, while each of the other three groups permutes the vertices of l transitively. Let δ be a 2-subspace in π_x containing l and let P_2 be the stabilizer of δ in $G(x)$. The following lemma is a direct consequence of Hypothesis 10.15.

Lemma 10.18 *If Hypothesis 10.15 holds then the quotient $P_2/G_1(x) \cong S_4$ acts faithfully on the six points in $\delta \cap \Gamma(x)$.* □

For $1 \leq i \leq 3$ define $P_1^{(i)}$ to be the subgroup of $G(l)$ generated by $G(x, l, \delta)$ together with the order-3 subgroup $S^{(i)}$, let $X^{(i)}$ be the subgroup in G generated by P_2 and $P_1^{(i)}$, and let $\Xi^{(i)}$ be the subgraph of Γ induced by the images of x under $X^{(i)}$.

Proposition 10.19 *The following assertions hold for every $1 \leq i \leq 3$:*

(i) *$\{P_1^{(i)}, P_2\}$ is a Goldschmidt amalgam;*
(ii) *$\Xi^{(i)}$ is a geometric subgraph in Γ at level 2;*
(iii) *$X^{(i)}$ acts faithfully on $\Xi^{(i)}$.*

Proof Let $B = P_1^{(i)} \cap P_2$ be the Borel subgroup. Then by the construction

$$[P_1^{(i)} : B] = [P_2 : B] = 3,$$

so (i) holds subject to the possibility of a non-trivial core, which will be eliminated in the proof of (iii). Then (ii) follows from (i), while (iii) is given by Lemma 10.18. □

Lemma 10.20 *Under Hypothesis* 10.15 *$G(x)$ is isomorphic to $2^3 \cdot L_3(2)$, i.e. to the unique non-split extension by $L_3(2)$ of its natural module.*

Proof First we claim that $|G(x)| = 2^6 \cdot 3 \cdot 7$. By Lemma 10.16 the number on the right-hand side of the equality is a lower bound for the order of $G(x)$. By Lemma 10.19, $G(x)$ contains with index 7 a parabolic subgroup P_2. The orders of the Borel subgroups in Goldschmidt's amalgam can be read from Tables 2.1 and 10.3. Therefore, in order to prove the claim it is sufficient to exclude the possibility that the amalgams $(P_1^{(i)}, P_2)$ are all isomorphic to G_5^1 for $1 \leq i \leq 3$, in which case $G_1(x)$ has order 2^4.

Let us take a closer look at that possibility. From Lemma 10.16 (iii) and the structure of the irreducible $L_3(2)$-modules over $GF(2)$, we conclude that, if $|G_1(x)| = 2^4$, then $G_1(x)$ involves one trivial one-dimensional module and a three-dimensional module (natural or dual natural). Since $G_1(x)$ acts on $\Gamma(y)$ as an elementary abelian group of order 4 on which $G(x, y)$ induces a fixed-point free action of the subgroup $S^{(0)}$ of order 3, we conclude that $G_1(x)/G_2(x)$ does not possess one-dimensional factor-modules and is isomorphic to the natural module of $L_3(2)$. Hence, the only option is that $G(x)$ contains a central involution. However, $P_2(G_5^1)$ contains a Sylow 2-subgroup of $G(x)$ but has no central involutions. This contradiction proves the claim.

It remains to prove that $G(x)$ does not split over $G_1(x)$. In fact, in the semidirect product $2^3 : L_3(2)$, in the stabilizer of a plane the largest normal abelian subgroup (of order 2^4) is elementary abelian (see Subsection 3.10.2), whereas in the parabolics $P_2(G_4^1)$ and $P_2(G_5)$ such a subgroup is isomorphic to $Z_4 \times Z_4$. □

Lemma 10.21 *In the above terms, two of the amalgams, say* $\{P_1^{(1)}, P_2\}$ *and* $\{P_1^{(2)}, P_2\}$, *are isomorphic to* G_4^1, *while the third one,* $\{P_1^{(3)}, P_2\}$, *is isomorphic to* G_5.

Proof By Lemma 10.20 we know that $G(x) \cong 2^3 \cdot L_3(2)$ and that $G(x, l) = C_{G(x)}(t)$ for some involution $t \in G_1(x) = O_2(G(x))$. Therefore,

$$E := G(x, l) \cong 2_+^{1+4}.S_3 \cong (Q_8 * Q_8).S_3.$$

The extraspecial group of order 2^5 of plus type possesses a unique presentation as a central product of two quaternion groups of order 8 (up to reordering the central factors). A Sylow 3-subgroup $S^{(0)}$ of E acts fixed-point freely on $O_2(E)/Z(E) \cong 2^4$. On the other hand,

$$\mathrm{Aut}(O_2(E)) \cong 2^4.(S_3 \times S_3).2 \cong 2^4.(S_3 \wr S_2)$$

has Sylow 3-subgroup T of order 3, which is elementary abelian. Among the other three subgroups, $S^{(1)}$, $S^{(2)}$ and $S^{(3)}$ of order 3 in T, two of them (say $S^{(1)}$ and $S^{(2)}$) centralize one of the Q_8-central factors, while the last one, $S^{(3)}$, acts on $O_2(E)/Z(E)$ fixed-point freely. Now the result follows from the remark after Table 2.1. □

Proposition 10.22 *Subject to Hypothesis* 10.15, *the locally projective amalgam is isomorphic to the amalgam associated with the action of* $G_2(3)$ *on Cooperstein's geometry* 4 *in Table 10.2.*

Proof By Lemma 10.20 we know the vertex stabilizer, while by Lemmas 10.19 and 10.21 we have two classes of geometric subgraphs $\Xi^{(1)}$ and $\Xi^{(2)}$ whose representatives possess foldings onto the generalized hexagon of $G_2(2)$. Then a reformulation of the proof of Corollary 3.4 in an article by Hoffman and Shpectorov[9] does the job. □

10.6.3 Outer Automorphism

The group $G_2(3)$ possesses an outer involutory automorphism α which cannot be realized as an automorphism of Cooperstein's geometry, since it permutes two classes $\Xi^{(1)}$ and $\Xi^{(2)}$ of planes. On the other hand, the automorphism can be realized as an automorphism α of the corresponding locally projective graph, leading to another locally projective action. Also α can be viewed as an automorphism of the amalgam from Proposition 10.22. In fact, if

$$(G(x), G(l)) \sim (2^3 \cdot L_3(2),\ 2_+^{1+4}.3^2.2)$$

and $F = \langle G, \alpha \rangle$, then

$$(F(x), F(l)) \sim (2^4 \cdot L_3(2),\ 2_+^{1+4}.(S_3 \times S_3)),$$

where $O_2(F(x))$ is the indecomposable extension of the natural $L_3(2)$-module by the trivial one-dimensional module. The argument in the previous subsection can be slightly modified to prove the following.

Proposition 10.23 *Suppose that the following conditions hold:*

(i) *G acts locally projectively on Γ of type* $(3, 3)$*;*
(ii) *for a vertex x we have $[G_{\frac{1}{2}}(x) : G_1(x)] = 2$, so that $G(x)$ acts on $\Gamma(x)$ as $L_3(2) \times 2$ acts on the 14 cosets of its 'diagonal' S_4-subgroup;*
(iii) $G_1(x) \neq 1$.

Then the corresponding projective amalgam is isomorphic to the one associated with action number 5 in Table 10.2.

Proof We can apply arguments as in the proof of Theorem 10.11 to recover the third class, $\Xi^{(3)}$, of geometric subgraphs (in this case they realize completions of the amalgam G_5^1). Then Lemma 10.18 and hence Proposition 10.19 (iii) hold. Therefore the Borel subgroup of the G_5^1-subamalgam is a Sylow 2-subgroup of the locally projective amalgam in question. Hence the

[9] C. Hoffman and S. Shpectorov, New geometric presentations for Aut $G_2(3)$ and $G_2(3)$, *Bull. Belg. Math. Soc.* **12** (2005), 813–826.

identification of the latter can be achieved along the same lines as in the proofs from the previous subsections. □

To end this section we would like to pose a problem which fits perfectly well the topic of the book.

Problem 10.24 *Construct a locally projective graph in which the stabilizer of a plane is isomorphic to M_{12} or to* $\mathrm{Aut}(M_{12})$.

10.7 Dual Space Graphs and Amalgams

In this section we consider the dual polar graph Γ on which the symplectic group $Sp_{2n}(2)$ acts locally projectively of type $(n, 3)$. We show explicitly how to describe the locally projective amalgam of type $(n, 2)$ corresponding to the densely embedded subgraph Δ stabilized by the orthogonal group $O^+_{2n}(2)$.

Let $V = V_{2n}(2)$ be a $2n$-dimensional $GF(2)$-space equipped with a non-singular symplectic form f. Let Γ be the *dual polar graph*, whose vertices are maximal (n-dimensional) totally singular subspaces in V with respect to f in which two vertex-subspaces are adjacent if their intersection is a hyperplane in each, and let $G \cong Sp_{2n}(2) = \mathrm{Aut}(V, f)$ be the corresponding symplectic group acting naturally on Γ.

Every edge of Γ is contained in a unique triangle in Γ, which is a *line*. Indeed, if X is a maximal totally singular subspace in V and U is a hyperplane in X then

$$U^\perp = U \oplus T,$$

where T is a *hyperbolic pair*, i.e. a non-singular two-dimensional subspace. Denoting the non-zero vectors of T by x, y, z, we obtain a line l in Γ consisting of the maximal totally singular subspaces

$$X = \langle U, x\rangle, \quad Y = \langle U, y\rangle \text{ and } Z = \langle U, z\rangle$$

containing U. The lines containing X are indexed by the hyperplanes of X, thus they carry the structure of the projective space π_X dual to X. The fact that $G(X)$ induces on X the full linear group

$$GL(X) \cong GL(X^*) \cong GL_n(2)$$

is implied by Witt's lemma.[10] Thus the action of G on Γ is locally projective of type $(n, 3)$ with the locally projective amalgam $\mathcal{A}(Sp_{2n}(2)) = \{G(X), G(l)\}$.

[10] P. Dembowski, *Finite Geometries*, Springer, Berlin, 1968.

Next we choose a quadratic form q of plus type whose associated bilinear form is f in the sense that

$$f(u,v) = q(u) + q(u+v) + q(v).$$

Let

$$H = \mathrm{Aut}((V,f,q)) \cong O_{2n}^{+}(2) < Sp_{2n}(2) \cong \mathrm{Aut}((V,f)) = G.$$

Assuming that X is totally singular with respect to q, we define Δ to be the subgraph in Γ induced by the images of X under H. In this case a vertex of Γ belongs to Δ precisely when it is a maximal totally isotropic subspace with respect to q, and Δ intersects the line l in an edge, say $\{X,Y\}$. Our goal is to describe the subamalgam $\mathcal{A}(O_{2n}^{+}(2)) = \{H(X), H(l)\}$ of $\mathcal{A}(Sp_{2n}(2))$ in terms independent of V, f and q. We start with the following well-known result.

Lemma 10.25 *The stabilizer $G(X)$ is a semidirect product with respect to the natural action of a subgroup $L_X \cong L_n(2)$ and an elementary abelian subgroup K_X of order $2^{n(n+1)/2}$, where $K_X = G_{\frac{1}{2}}(X)$. Furthermore, the subgroup $G_1(X)$ is an L_X-submodule N_X in K_X that is isomorphic to the exterior square of X, K_X/N_X is the dual module X^* of X, and K_X does not split over N_X.*

Proof Since the order of G is known to be

$$|G| = |Sp_{2n}(2)| = 2^{n^2} \cdot \prod_{i=1}^{n} (2^{2i}-1),$$

while the number of vertices in Γ is also known, in order to prove the lemma we need to produce the required automorphisms as stated in the lemma. The subgroup L_X can be obtained as $G(X) \cap G(X^{\mathrm{o}})$, where X^{o} is a maximal totally singular subspace disjoint from X. The form f establishes a duality δ between X and X^{o} via

$$\delta : v \mapsto X^{\mathrm{o}} \cap v^{\perp},$$

where $v \in X$.

On the other hand, the whole of f can be recovered from the triple $(X, X^{\mathrm{o}}, \delta)$ by declaring that X and X^{o} are totally singular and disjoint, while for $v \in X$, $u \in X^{\mathrm{o}}$ the equality $f(v,u) = 0$ holds if and only if $u \in \delta(v)$. Then we put

$$L_X = \mathrm{Aut}(X, X^{\mathrm{o}}, \delta) \cong L_n(2).$$

The subgroup K_X is generated by the *symplectic transvections*

$$t_v : u \mapsto u + f(u,v)v$$

taken for all $v \in X$. It is clear that t_v is an involution, which stabilizes X vectorwise, and that $[t_v, t_w] = 1$ for all $v, w \in X$. If $v, w, \in T$ then the product of t_v, t_w and t_{v+w} is a *Siegel transformation*

$$s_R := u \mapsto u + f(u, v)v + f(u, w)w.$$

On the other hand, if 2-subspaces R and Q in X generate a 3-subspace, then $s_R s_Q = s_{R+Q}$, which means that K_X is the $L_n(2)$-permutation module on the set of non-zero vectors of X, the quotient over the submodule spanned by the sums of the vectors in 3-subspaces. This demonstrates that K_X is precisely as stated in the lemma, where N_X is spanned by the Siegel transformations associated with 2-subspaces in X. □

The stabilizer $G(l)$ of a line is generated by $G(x, l)$ together with the Levi S_3-subgroup

$$S_3^{(l)} = \langle t_x, t_y, t_z \rangle,$$

while $O_2(G(x, l))$ is complemented in $G(x, l)$ by a Levi $L_X(l) \cong L_{n-1}(2)$-subgroup. These are contained in the Levi subgroup which is the stabilizer of the orthogonal decomposition

$$U^\perp = U \oplus \langle x, y, z \rangle.$$

Now we involve the quadratic form q of plus type whose associated bilinear form is f. Having f in place and assuming that both X and its opposite X^{o} are totally singular with respect to q, we determine q uniquely. In particular $L_X \cong L_n(2)$ preserves q. It is well known and easy to check that the symplectic transvection t_u preserves q if and only if u is non-singular and the Siegel transformation s_R preserves q whenever R is totally isotropic. This immediately gives the following.

Lemma 10.26 *The stabilizer of X in $H \cong O_{2n}(2)^+$ is a semidirect product of L_X and*

$$G_1(X) = N_X \cong X \wedge X \cong 2^{n(n-1)}. \qquad \square$$

We assume without loss of generality that $\delta(U) \cap X^{\mathrm{o}} = \langle y \rangle$. In this case $O_2(L_X(U)) \cong 2^{n-1}$ is generated by the Siegel transformations $s(u, y)$ taken for all $u \in U$. This gives the following.

Lemma 10.27 *The subgroup*

$$R_X := \langle s(u, y) \mid u \in U \rangle$$

induces on $\Gamma(x)$ an elementary abelian action of order $2^{(n-1)}$ which coincides with the action of $G_1(y)$ on that set. □

On the other hand, it is easy to see that the subgroup

$$R_Y := \langle s(u,x) \mid u \in U \rangle$$

induces on $\Gamma(y)$ an elementary abelian action of order $2^{(n-1)}$ which coincides with the action of $G_1(y)$ on that set. This leads to the principal assertion given in the following proposition.

Proposition 10.28 *Define the subgroups $H(x)$ and $H(l)$ in $G(x)$ and $G(l)$, respectively, by the following rules:*

(i) $H(X) = \langle N_X, L_X \rangle$;
(ii) $H(l) = \langle G_1(l), R_X, R_Y, t_z, L_X(l) \rangle$.

Then $H := \langle H(X), H(l) \rangle \cong O_{2n}(2)^+$ is the stabilizer of the densely embedded subgraph Δ in Γ associated with q and $(H(X), H(l))$ is the locally projective amalgam associated with the action of H on Δ. □

10.8 Towards the Planes of Symplectic Type

In this section, subject to exclusion of the situations

(A) $[G_{\frac{1}{2}}(x) : G_1(x)] \leq 2$ (accomplished in Proposition 10.23) and
(B) $G_1(x) = 1$,

we intend to show that $G(x)$ induces on $\Gamma(x)$ a split extension by $L_3(2)$ of its dual natural module,

$$G(x)/G_1(x) \cong V_3^* : L_3(2),$$

and that every plane realizes a completion of Goldschmidt's amalgam G_3^1 which possesses a completion

$$S_6 \cong Sp_4(2),$$

whence the title of the section. In view of the inductive nature of our argument throughout the section we assume that $n = 3$.

10.8.1 Some $L_3(2)$-Modules and Extensions

In this subsection we collect some well-known (and some less well-known) facts about $GF(2)$-representations of $L_3(2)$ as in Lemma 6.22.

Lemma 10.29 *Let P_7 be the $GF(2)$-permutation module of $L_3(3)$ acting on the set of non-zero vectors on V_3. Then*

(i) $P_7 = (V_3^*/V_3) \oplus V_1$, *that is a direct sum of the indecomposable extension of* V_3^* *by* V_3 *and the trivial one-dimensional module;*
(ii) *all indecomposable extensions of* V_3^* *by* V_3 *are isomorphic;*
(iii) *an extension of* V_3^*/V_3 *by* $L_3(2)$ *modulo* V_3^* *splits over* V_3*;*
(iv) *the semidirect product* $P_7 : L_3(2)$ *with respect to the natural action contains subgroups of the form* $V_3^*.L_3(2)$ *(both split and non-split).*

Proof The item (iii), included for completeness, has been checked by Derek Holt using his cohomology package.[11] □

Lemma 10.30 *Let* U *be the quotient of the* $GF(2)$*-permutation module* P_{21} *on the flags* (p, l) *of the Fano plane subject to the relations*

$$(p_1, l) + (p_2, l) + (p_3, l) = 0 \text{ whenever } l = \{p_1, p_2, p_3\}$$

and

$$(p, l_1) + (p, l_2) + (p, l_3) = 0 \text{ whenever } l_1, l_2, l_3 \text{ are the lines through } p.$$

Then $U = V_8 \oplus V_1$.

Proof The assertion is in fact one of the definitions of the Steinberg module. □

10.8.2 Eliminating a Direct Factor

We continue to assume that $n = 3$. Our aim is to analyse the possibilities for the action of $G(x)$ on $\Gamma(x)$. The subgraph induced by $\Gamma(x)$ is the disjoint union of seven edges denoted by $7K_2$ and this set serves as a point-set of a projective plane π_x associated with x. Therefore, the action induced by $G(x)$ on $\Gamma(x)$ is a subgroup of

$$A := \text{Aut}(7\,K_2, \pi_x) \cong P_7 : L_3(2).$$

The submodule structure of P_7 is described in Lemma 10.29 (i).

The first reduction is the following.

Lemma 10.31 *Suppose that* $G(x)$ *acting on* $\Gamma(x)$ *contains a direct summand, which is the trivial one-dimensional module. Then the corresponding locally projective amalgam* $\{G(x), G(l)\}$ *contains an index-2 subamalgam* $\{K(x), K(l)\}$ *which is also locally projective of type* $(n, 3)$*. More specifically, there is an automorphism* α_x *of* $K(x)$ *and an automorphism* α_l *of* $K(l)$ *whose*

[11] D. Holt, private communication, 2017.

squares are inner automorphisms and which coincide on the intersection $K(x) \cap K(l)$, *and*

$$G(x) = \langle K(x), \alpha_x \rangle, \quad G(l) = \langle K(l), \alpha_l \rangle.$$

Proof The non-identity element from the direct summand swaps every edge in the subgraph induced by $\Gamma(x)$. Therefore, if Ξ is a plane (which is a geometric subgraph at level 2) passing through x with stabilizer F, then $G(x) \cap F$ acts on $\Xi(x)$ as $S_4 \times 2$. As indicated in Table 10.3, $G(x) \cap F$ induces on Ξ either $P_1(G_4^1)$ or $P_2(G_5^1)$. Both G_4^1 and G_5^1 are not 2-perfect. Therefore we define K to be the subgroup in G generated by the commutator (index-2) subgroup of $G(x)$ and by the index-2 subgroup of the universal completion of G_4^1 or G_5^1, to obtain G_4 or G_5, respectively.

We assume that Γ is a bipartite half of the tree, the latter being the coset graph of the universal completion of $\{G(x), G(l)\}$). Then K is an index-2 subgroup in G and its stabilizers $K(x)$ and $K(l)$ are as described in the lemma. It is worth mentioning that G is generated by $G(x)$ together with the stabilizer of the plane Ξ. It is clear that the direct summand of order 2 is not contained in $K(x)$ and hence the required reduction is accomplished. □

10.8.3 Further Reduction

The case in which the action of $G(x)$ on $\Gamma(x)$ involves the dual natural module V_3 appears a bit harder to handle, and we formulate it as follows.

Problem 10.32 Classify the locally projective graphs of type $(n, 3)$, such that

(i) $G(x)$ induces on $\Gamma(x)$ the group $V_3^*/V_3.L_3(2)$;
(ii) the stabilizer of a plane induces on that plane a completion of Goldschmidt's amalgam G_4^1 or G_5^1.

For the remaining part of the section we work under the following hypothesis.

Hypothesis 10.33 (Γ, G) *is a locally projective pair of type* $(3, 3)$, *and for a vertex* x *of* Γ *we have*

$$G(x)^{\Gamma(x)} \cong G(x)/G_1(x) \cong V_3^*.L_3(2),$$

where the latter extension splits.

For the remainder of the section we will work through the first column of Table 10.3, leaving the top entry G_3^1 alone, and try to eliminate the remaining possibilities subject to Hypothesis 10.33. Recall that the subgroups in the first column describe the action on a plane Ξ passing through x of the subgroup which stabilizes both x and Ξ. The aim is to show that the reconstruction of the action of $G(x)$ on the ball with radius 3 centred at x using these data leads to a contradiction.

$\underline{P_1(G_4)}$ Under Hypothesis 10.33, $O_2(G(x, y))$ induces on $\Gamma(y)$ an elementary abelian group of order 2^4, which contradicts the structure of $M_1^2 \cong Z_4$.

$\underline{P_2(G_4)}$ By calculating with characters of $G_2(2)'$ or otherwise, we deduce that $P_4(G_4)$ contains 15 involutions. Three of them correspond to the lines passing through x and are a^2, b^2 and a^2b^2.

The remaining 12 involutions correspond to the lines at distance 1 from x. Each of these 12 involutions normalizes a subgroup of order 3, which acts fixed-point freely on the set of triangles passing through x. Therefore an involution i from the 12-orbit stabilizes one line through x and swaps the other two.

Upon reversing the picture, we observe that $G_1(x) \cap G(\Xi) = \langle a^2, b^2, a^2b^2 \rangle$ is not contained in $G_{\frac{3}{2}}(x)$. On the other hand, $O_2(G(x, y))/G_1(x)$ contains two S_3-invariant subgroups outside $G_{\frac{1}{2}}(x)$. This enables us to describe the action of $G_1(x)$ on $\Gamma_2(x)$: it acts with 21 orbits, inducing an order-2 action on each. The orbits are indexed by the flags of π_x and satisfy the same relations as in Lemma 10.30. Hence $G_1(x)/G_2(x)$ is the Steinberg module of $L_3(2)$ (possibly together with a one-dimensional trivial direct factor).

Now, since a^2, b^2 and a^2b^2 are squares of elements a, b and ab of order 4 acting as non-trivial elements of $G_{\frac{1}{2}}(x)$, there must be a non-trivial squaring map from $G_{\frac{1}{2}}(x)/G_1(x)$ to $G_1(x)/G_2(x)$. But clearly it is impossible to construct such a squaring map from the dual natural module of $L_3(2)$ into the Steinberg module. Thus we have encountered a contradiction.

$\underline{P_1(G_4^1)}$ Since $M_0^1 \cong S_4 \times 2$, this case falls under Problem 10.32.

$\underline{P_1(G_5)}$ In this case we consider the submodule in $G_1(x)/G_2(x)$ spanned by the images of the 2-subspace $\langle a^2, b^2, a^2b^2 \rangle$. Arguing as in the case of $P_2(G_4)$, we identify this submodule with the Steinberg module. The non-existence of the squaring mapping from the dual natural $L_3(2)$-module to the Steinberg module is the way to reach the contradiction.

$\underline{P_1(G_4^1) \text{ and } P_1(G_5^1)}$ These two cases are analogous. Here we are also able to deduce that $G_1(x)/G_2(x)$ contains a Steinberg module, although there is no necessity for square mapping inside, so $G_{\frac{1}{2}}(x)/G_2(x)$ may well be elementary abelian. Since the Steinberg module is projective, we have

$$G_{\frac{1}{2}}(x)/G_2(x) \cong V_3^* \oplus V_8.$$

Considering the images of M_3^2 under $G(x)$ acting by conjugation, we conclude that they generate a non-trivial quotient Q of the $L_3(2)$-permutation module on the lines of π_x. The quaternion subgroups in the parabolic $P_1(G_5^{(1)})$ force a squaring map of V_3^* into Q, which it is impossible to construct whilst satisfying the local constraints from the parabolics.

$\underline{P_2(G_5^1)}$ Since $M_0^1 \cong S_4 \times 2$, this case also falls under Problem 10.32.

$\underline{P_2(G_4^1)}$ This can be reduced to the $P_2(G_4)$ case by considering the subgroup in $G(x)$ generated by the index-2 subgroups of the seven universal completions of G_4^1, which are the actions on the seven planes passing through x. Then the actions become the universal completions of G_4 and the reduction works.

$\underline{P_2(G_5^1)}$ Since $M_0^1 \cong S_4 \times 2$, this case falls under Problem 10.32.

Therefore, we have accomplished the reduction. So, leaving out the case of faithful action of $G(x)$ on $\Gamma(x)$, we have the following.

Proposition 10.34 *One of the following holds, where G is a group acting locally projectively of a graph Γ of type $(m, 3)$ for $m \geq 3$ with $G_1(x) \neq 1$.*

(ii) *Γ contains a geometric subgraph at level 3 on which the action of its stabilizer falls under the conditions in Problem 10.32;*

(ii) (a) *$G(x)$ induces in $\Gamma(x)$ a semidirect product $2^n : L_n(2)$;*

(b) *the stabilizer of a plane induces on that plane a completion of Goldschmidt's amalgam G_3^1.*

Among the examples in Table 10.2, the following ones satisfy the condition in Proposition 10.34 (ii):

2a, 3a, 6a, 7a, 8a, 10a.

They include the the dual polar space graphs of $Sp_{2m}(2)$, the recently discovered example completing in A_{16} and the famous Mathieu–Conway Monster sequence of tilde geometries. This again demonstrates the role of the Mathieu groups, as a path towards larger sporadic simple groups. Each of the above-mentioned examples contain a densely embedded subgraph:

2b, 3b, 6b, 7b, 8b, 10b,

respectively.

10.9 Constructing Densely Embedded Subgraphs

We follow the construction procedure outlined in (DE1) to (DE4) in Section 10.3. The construction of $H(x)$ goes through because of Proposition 10.34 (ii) (a) and the corresponding $L_n(2)$-complement will be denoted by L_x. The next lemma goes back to discussions in Chapter 4.

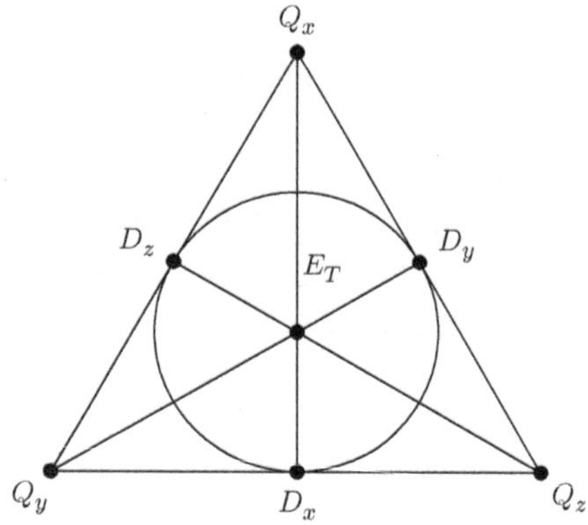

Figure 10.1

We analyze the action of $A_T = G(x, y.z)$ on $\Gamma(l)$, where as always $l = \{x, y, z\}$ is a line containing x and the factorization over

$$G_1(l) := G_1(x) \cap G_1(y) \cap G_1(z)$$

is implicit.

Lemma 10.35 *The following assertions hold:*

(i) *the image of $L_x \cap G_1(l)$ is a semidirect product of $L_x(l) \cong L_{n-1}(2)$ and a dual natural module of that group which we denote by $R_x(l)$;*
(ii) *$O_2(A_T)$ is elementary abelian of order 2^9; it is generated by $Q_u = G_1(u)/G_1(u)$ for $u \in l$ as shown in Figure 10.1 and contains seven copies of the dual natural module of $L_x(l)$;*
(iii) *D_x is the image of $G_{\frac{1}{2}}(x)$ in A_T;*
(iv) *up to the choice between y and z, we may assume that $R_x(l) = Q_y$.*

Proof The result follows directly from the local properties of the plane stabilizer, which can already be checked inside the S_6-completion of the G_3^1-amalgam. □

Now the action of σ in $O_2(A_T)$ is clear: it must swap $Q_y = R_x(l)$ and Q_x. The big question is how to choose σ to keep $L_x(l)$ inside $H(x)$. The question can be solved inside A_T (see Section 6.1) but one definitely requires additional information on the starting locally projective amalgam $\{G(x), G(l)\}$.

When completing the classification one should also address the case in which the action of $G(x)$ on $\Gamma(x)$ is faithful. This would include the projective space example (9b in Table 10.2) together with the A_7-example, which has already been characterized in a different context, and the whole variety of Hamming-type examples 11a. Finally, it should be checked that the example 6a, which constitutes the central topic of the present book, is unique of its kind.

References

M. Aschbacher, *Sporadic Groups*, Cambridge University Press, Cambridge, 1994. (Cited on page 6.)

M. Aschbacher, *Finite Group Theory*, Cambridge University Press, Cambridge, 2000. (Cited on page 93.)

M. Aschbacher and Y. Segev, Extending morphisms of groups and graphs, *Ann. Math.* **135** (1992), 297–323. (Cited on page 11.)

A. E. Brouwer, The uniqueness of the near hexagon on 759 points, in *Finite Geometries*, ed. N. L. Johnson, M. J. Kallaher and C. T. Long, Marcel Dekker, New York, 1982, pp. 47–60. (Cited on page 6.)

A. E. Brouwer, A. M. Cohen and A. Neumeier, *Distance Regular Graphs*, Springer, Berlin, 1989, p. 386. (Cited on page 102.)

W. Burnside, *Theory of Groups of Finite Order*, second edition, Cambridge University Press, Cambridge, 1911. (Cited on page 108.)

C. Choi, On subgroups of M_{24}. I: Stabilizers of subsets, *Trans. Amer. Math. Soc.* **167** (1972), 1–27. (Cited on page 77.)

C. Choi, On subgroups of M_{24}. II: The maximal subgroups of M_{24}. *Trans. Amer. Math. Soc.* **167** (1972), 29–47. (Cited on page 77.)

J. H. Conway, Three lectures on exceptional groups, in *Finite Simple Groups*, ed. M. B. Powell and G. Higman, Academic Press, New York, 1971, pp. 215–247. (Cited on page 23.)

J. H. Conway, The Golay Codes and the Mathieu Groups, in *Sphere Packings, Lattices and Groups*, ed. J. Conway and N. J. A. Sloane, Springer, New York, 1988, pp. 299–330. (Cited on page xi.)

A. Delgado, D. Goldschmidt and B. Stellmacher, *Groups and Graphs: New Results and Methods*, Birkhäuser, Basel, 1985. (Cited on page 11.)

P. Dembowski, *Finite Geometries*, Springer, Berlin, 1968. (Cited on page 83, 156.)

L. E. Dickson, *Linear Groups: With an Exposition of the Galois Field Theory*, Dover Publications, New York, 1958. (Cited on page 90.)

J. D. Dixon and B. Mortimer, *Permutation Groups*, Springer, Berlin, 1996. (Cited on page 6.)

D. Z. Djoković and G. L. Miller, Regular groups of automorphisms of cubic graphs, *J. Combin. Theory* (B) **29** (1980), 195–230. (Cited on page 141.)

T. Eguchi, H. Ooguri and Y. Tachikawa, Notes on the $K3$ surface and the Mathieu group M_{24}. *Exper. Math.* **20** (2011), 91–96. (Cited on page 105, 139.)

G. Frobenius, Über die Charaktere der mehrfach transitiven Gruppen, *Sitzungsber. Königl. Preuß. Akad. Berlin* (1904), 558–571. (Cited on page 105.)

T. Gannon, Much ado about Mathieu, *Adv. Math.* **301** (2016), 322–358. (Cited on page 139.)

M. J. E. Golay, Notes on digital coding, *Proc. IRE* **37** (1949), 657. (Cited on page 6.)

D. M. Goldschmidt, Automorphisms of trivalent graphs, *Ann. Math.* **111** (1980), 377–406. (Cited on page 10, 147.)

D. Gorenstein, *Finite Simple Groups. An Introduction to Their Classification*, Springer, Berlin, 1982. (Cited on page 133.)

R. L. Griess, The friendly giant, *Invent. Math.* **69** (1982), 1–102. (Cited on page 133.)

M. Hall, Note on the Mathieu group M_{12}, *Arch. Math.* **13** (1962), 334–340. (Cited on page 88.)

D. Held, The simple groups related to M_{24}, *J. Algebra* **13** (1969), 253–296. (Cited on page 120, 121.)

C. Hoffman and S. Shpectorov, New geometric presentations for Aut$G_2(3)$ and $G_2(3)$, *Bull. Belg. Math. Soc.* **12** (2005), 813–826. (Cited on page 155.)

A. A. Ivanov, *Geometry of Sporadic Groups* I, Cambridge University Press, Cambridge, 1999. (Cited on page xi, 6, 96.)

A. A. Ivanov, Y-groups via transitive extensions, *J. Algebra* **218** (1999), 412–435. (Cited on page xi.)

A. A. Ivanov, J_4, Oxford University Press, Oxford, 2004. (Cited on page xi, 117.)

A. A. Ivanov, *The Monster Group and Majorana Involutions*, Cambridge University Press, Cambridge, 2009. (Cited on page xi, 117.)

A. A. Ivanov and U. Meierfrankenfeld, A computer-free construction of J_4, *J. Algebra* **219** (1999), 113–172. (Cited on page 105.)

A. A. Ivanov and S. V. Shpectorov, A geometry for the O'Nan–Sims group connected to the Petersen graph, *Russian Math. Surveys* **41** (1986), 211–212. (Cited on page 133.)

A. A. Ivanov and S. V. Shpectorov, The P-geometry for M_{23} has no nontrivial 2-coverings, *Europ. J. Combin.* **10** (1990), 347–362. (Cited on page 79.)

A. A. Ivanov and S. V. Shpectorov, Applications of group amalgams to algebraic graph theory, in *Investigation in Algebraic Theory of Combinatorial Objects*, ed. I. A. Faradzev, A. A. Ivanov, M. H. Klin and A. J. Woldar, Kluwer, Dordrecht, 1994, pp. 417–442. (Cited on page 11.)

A. A. Ivanov and S. V. Shpectorov, *Geometry of Sporadic Groups* II, Cambridge University Press, Cambridge, 2002 (Cited on page xi.)

A. A. Ivanov and S. V. Shpectorov, Amalgams determined by locally projective actions, *Nagoya Math. J.* **176** (2004), 19–98. (Cited on page 142.)

W. M. Kantor, Automorphism groups of Hadamard matrices, *J. Comb. Theory* **6** (1969), 279–281. (Cited on page 89.)

S. Kondo, Niemeier lattices, Mathieu groups, and finite groups of symplectic automorphisms of $K3$ surfaces, *Duke. Math. J.* **92** (1998), 593–603. (Cited on page 137.)

J. H. van Lint, *Introduction to Coding Theory*, third edition, Springer, Berlin, 1999. (Cited on page 95.)

É. Mathieu, Mémoire sur l'étude des fonctions de plusieurs quantités sur la manière de les former et sur les substitutions qui les laissent invariable. *J. Math. Pure Appl.*, sér. II, **6** (1861), 241–328. (Cited on page 86.)

É. Mathieu, Sur la fonction cinq fois transitive de 24 quantités, *J. Math. Pure Appl.* **18** (1873), 25–46. (Cited on page 6.)

S. Mukai, Finite groups of automorphisms of $K3$ surfaces and the Mathieu group, *Invent. Math.* **94** (1988), 183–221. (Cited on page 137.)

V. V. Nikulin, Konechnye gruppy avtomorfizmov kelerovykh poverkhostei tipa $K3$ [Finite groups of automorphisms of Kählerian surfaces of type $K3$], *Trudy Mosk. Mat. Ob.* [*Trans. Moscow Math. Soc.*] **38** (1980), 71–137. (Cited on page 137.)

M. E. O'Nan, Some evidence for the existence of a new simple group, *Proc. London Math. Soc.* **32** (1976), 421–479. (Cited on page 133.)

C. Parker and P. Rowley, On the non-commuting case for $(S_3, L_3(2))$-amalgams, *J. Algebra* **181** (1996), 267–285. (Cited on page 143.)

A. Pasini, Some remarks on covers and apartments, in *Finite Geometries*, ed. C. A. Baker and L. M. Batten, Marcel Dekker, New York, 1985, 233–250. (Cited on page 9.)

A. Pasini, *Diagram Geometries*, Oxford University Press, Oxford, 2006. (Cited on page 76.)

S. Rees and L. H. Soicher, An algorithmic approach to fundamental group and covers of combinatorial cell complexes, *J. Symb. Comp.* **29** (2000), 59–77. (Cited on page 13.)

M. A. Ronan, Covering and automorphisms of chamber systems, *Europ. J. Comb.* **1** (1980), 259–269. (Cited on page 13.)

M. A. Ronan, Locally truncated buildings and M_{24}, *Math. Z.* **180** (1982), 469–501. (Cited on page 7, 73.)

M. A. Ronan and S. D. Smith, 2-Local geometries for some sporadic groups, in *Proceedings of Symposia in Pure Mathematics 37 (Finite Groups)*, American Mathematical Society, 1980, pp. 283–289. (Cited on page 71.)

M. A. Ronan and G. Stroth, Minimal parabolic geometries for the sporadic groups, *Europ. J. Combin.* **5** (1984), 59–91. (Cited on page 74.)

J.-P. Serre, *Arbres, amalgams, SL_2*, Astérisque 46, Société Mathématique de France, Paris, 1977. (Cited on page 116.)

J. P. Serre, *Trees*, Springer, New York, 1980. (Cited on page 11.)

E. E. Shult, lecture notes on coverings of graphs, Kansas State University, 1997. (Cited on page xi.)

J. G. Thompson, Finite dimensional representations of free products with an amalgamated subgroup, *J. Algebra* **69** (1981), 146–149. (Cited on page 14.)

J. Tits, Ensembles ordonnés, immeubles et sommes amalgamées, *Bull. Soc. Math. Belg.* **A38** (1986), 367–387. (Cited on page 9.)

V. I. Trofimov, Vertex stabilizers of locally projective groups of automorphisms of graphs: A summary, in *Groups, Combinatorics and Geometry*, ed. A. A. Ivanov, M. W. Liebeck and J. Saxl, World Scientific, River Edge, NJ, 2003, pp. 313–326 (Cited on page 142.)

W. T. Tutte, A family of cubical graphs, *Proc. Camb. Phil. Soc.* **43** (1947), 459–474. (Cited on page 141.)

R. Weiss, Elations of graphs, *Acta Math. Acad. Scient. Hungary* **34** (1979), 101–103. (Cited on page 146.)

E. Witt, Über Steinersche Systeme, *Abh. Math. Seminar Hamburg* **12** (1938), 265–275. (Cited on page 6.)

Index

For EU product safety concerns, contact us at Calle de José Abascal, 56–1°, 28003 Madrid, Spain or eugpsr@cambridge.org.

www.ingramcontent.com/pod-product-compliance
Lightning Source LLC
LaVergne TN
LVHW101502200926
848483LV00001B/202